AF782795

The Ecological Imaginary in Literature and Other Media

From Gilgamesh to Gawain and the Green Knight, the Brothers Grimm to Grimdark, the natural world has provided the backdrop for Fantasy since its earliest iterations. The playgrounds of childhood are often a writer's first Fantasy landscape and can develop into fully fledged storyworlds. Do readers of Fantasy seek out the genre for a taste of this unsullied environment? Is it nostalgia for the lost Edens of childhood, a way to escape, or to find resilience and inspiration? And in a time of Climate Emergency, is the nature of Fantasy changing to reflect the challenges *it* presents? Can the blue-sky thinking of the Fantastic provide us with a useful tool for addressing what the United Nations has called 'the defining crisis of our time'? This is a timely survey of the environmental aspects of Fantasy, with a unique focus on Fantasy sites and the real-world impact of Fantasy texts across media.

Kevan Manwaring, Editor of the *British Fantasy Society Journal*, is the MA Creative Writing programme leader at Arts University Bournemouth. He is the author of *Writing Ecofiction, Desiring Dragons, The Windsmith Elegy, Ballad Tales*, and editor of *Heavy Weather*. A contributor to *Free Thinking* on BBC Radio 3 (panels on Vikings, Ursula K. Le Guin, and John Cowper Powys), he was the Academic Consultant for BBC 4's *The Secret Life of Books*. He is a member of the Climate Fiction Writers League.

Routledge Studies in Speculative Fiction

Contemporary Pakistani Speculative Fiction and the Global Imaginary
Democratizing Human Futures
Shazia Sadaf and Aroosa Kanwal

J.R.R. Tolkien in Central Europe
Context, Directions, and the Legacy
Edited by Janka Kascakova and David Levente Palatinus

Rethinking Gothic Transgressions of Gender and Sexuality
New Directions in Gothic Studies
Edited by Sarah Faber and Kerstin-Anja Münderlein

Fictional Languages in Science Fiction Literature
Stylistic Explorations
Israel A. C. Noletto

Caribbean Futurism and Beyond
Conversations with writers of Folklore, Fantasy, Science, and Speculative Fiction
Jarrel De Matas

Magical Realism in Twenty-First-Century Transnational Fiction
The Global Supernatural
D. Chichester

The Ecological Imaginary in Literature and Other Media
The Nature of Fantasy
Kevan Manwaring

For more information about this series, please visit: https://www.routledge.com/Routledge-Studies-in-Speculative-Fiction/book-series/RSSF

The Ecological Imaginary in Literature and Other Media

The Nature of Fantasy

Kevan Manwaring

Routledge
Taylor & Francis Group
NEW YORK AND LONDON

First published 2026
by Routledge
605 Third Avenue, New York, NY 10158

and by Routledge
4 Park Square, Milton Park, Abingdon, Oxon, OX14 4RN

Routledge is an imprint of the Taylor & Francis Group, an informa business

© 2026 Kevan Manwaring

The right of Kevan Manwaring to be identified as author of this work has been asserted in accordance with sections 77 and 78 of the Copyright, Designs and Patents Act 1988.

All rights reserved. No part of this book may be reprinted or reproduced or utilised in any form or by any electronic, mechanical, or other means, now known or hereafter invented, including photocopying and recording, or in any information storage or retrieval system, without permission in writing from the publishers.

Trademark notice: Product or corporate names may be trademarks or registered trademarks, and are used only for identification and explanation without intent to infringe.

Library of Congress Cataloging-in-Publication Data
Names: Manwaring, Kevan author
Title: The ecological imaginary in literature and other media : the nature of fantasy / Kevan Manwaring.
Description: New York, NY : Routledge, 2026. | Series: Routledge studies in speculative fiction | Includes bibliographical references and index.
Identifiers: LCCN 2025023765 (print) | LCCN 2025023766 (ebook) | ISBN 9781032733210 hardback | ISBN 9781032733227 paperback | ISBN 9781003463665 ebook
Subjects: LCSH: Fantasy fiction—History and criticism | Ecocriticism | Fantasy in mass media | LCGFT: Literary criticism
Classification: LCC PN3435 .M34 2026 (print) | LCC PN3435 (ebook) | DDC 809.3/8766—dc23/eng/20250617
LC record available at https://lccn.loc.gov/2025023765
LC ebook record available at https://lccn.loc.gov/2025023766

ISBN: 978-1-032-73321-0 (hbk)
ISBN: 978-1-032-73322-7 (pbk)
ISBN: 978-1-003-46366-5 (ebk)

DOI: 10.4324/9781003463665

Typeset in Sabon
by Apex CoVantage, LLC

Contents

Introduction: Fantastical Meditations 1

1 Early Fantasy: Green Knights and Faerie Queens 28

2 Children's Fantasy: The Mythopoeic Landscapes of Childhood 52

3 Epic Fantasy: Portals, Paradises, and Wastelands 80

4 Urban Fantasy: The Re-Enchantment of the City 102

5 21st-Century Fantasy—Fantastic Ecologies in the Capitalocene 124

6 Mapping the Fantastic: The Ethics of Placiality, Architectonics, and Speculative Cartography 152

7 Fantasy Sites: Fantasy Texts and Their Real-World Impact 174

Conclusion: The Le Guin Index and Other Interventions 192

Bibliography *201*
Index *219*

Introduction

Fantastical Meditations

Hills of Vision

Through its multifarious effects on the sensory system and consciousness of those who enter it, nature—in the sense of a predominantly bio-organic, rather than artificial, environment—has been inspiring writers for a long time (Schama, 1995; Owens, 2021). And, so likewise, the cultivation of Fantasy, in its most generous definition to include all works that imaginatively engage with the porousness of reality, rather than slavishly replicate its mere materiality. Time in the natural world encourages an expansion of awareness into the more-than-human and beyond. A walk to a viewpoint can catalyse blue-sky thinking as we experience a perspectival shift, our horizons widen, and our gaze travels further. Nature facilitates consciousness shifts from the microscopic to the telescopic, and vice versa. This altitudinal thinking is both facilitated by nature and mirrored in the fluid ontological topographies of Fantasy.

In the 14th-century Christian allegorical poem *Piers Plowman*, written chiefly in 1380 during the wake of the Black Death as a critique on the waywardness of society and its way back to salvation (via the no-nonsense dogma of the titular farmer, whom the main seeks, meeting along the way a cross section of Medieval society), the protagonist, Will (a loosely fictionalised William Langland), receives a vision while sleeping on the Malvern Hills (the particular hill of vision being identified as British Camp, an Iron Age hillfort atop Herefordshire Beacon):

> In a summer season, when soft was the sunshine,
> In shrouds did I shape me, like as a poor shepherd;
> In habit as an hermit, unholy of living,
> Went wide in this world, wonders to hear of.
> But on a May morning, on a Malvern hillside,
> A marvel I met with, of magical semblance.
> I was weary with wandering, and went for to rest me

DOI: 10.4324/9781003463665-1

> Under a broad bankside, by a brooklet's margin;
> And as I lay and leaned there, and looked into the waters,
> I slumbered in a sleeping, it sounded so merry.
>
> Thereon fell I to dreaming a dream full of marvels:
> That I was in a wilderness—where, never wist I.
> Then as I held my eyes eastward, on high where the sun was,
> I saw a keep on crest, cunningly builded.
> A deep dale was below, a dungeon within it,
> With deep ditches and dark and dreadful appearance.
> A fair field full of folk I found there between them,
> Of every manner of men, the mean and the wealthy,
> Working and wandering, as the way of the world is.
> (trans. From Spencer, ed., 1965: 68)

Such a scene, remediated into prose fiction, would not be out of place in many a Fantasy novel. It contains several key tropes endlessly reworked in the 'iterations of Fantasy' (Sangster, 2023): the lonely wanderer, the vision-inducing hill, wilderness, a castle, dungeons, the stratification of society from the peasant class to the privileged. Will could be Frodo Baggins, sitting in the 'Seat of Seeing' on Amon Hen, beholding a terrifying overview of Middle-Earth and all of its fault-lines:

> Up he went and sat upon the ancient chair, feeling like a lost child that had clambered upon the throne of mountain-kings. At first he could see little. He seemed to be in a world of mist in which there were only shadows: the Ring was upon him. Then here and there the mist gave way and he saw many visions: small and clear as if they were under his eyes upon a table, and yet remote. There was no sound, only bright living images. The world seemed to have shrunk and fallen silent. He was sitting upon the Seat of Seeing, on Amon Hen, the Hill of the Eye of the Men of Númenor. Eastward he looked into wide uncharted lands, nameless plains, and forests unexplored. Northward he looked, and the Great River lay like a ribbon beneath him, and the Misty Mountains stood small and hard as broken teeth. Westward he looked and saw the broad pastures of Rohan; and Orthanc, the pinnacle of Isengard, like a black spike. Southward he looked, and below his very feet the Great River curled like a toppling wave and plunged over the falls of Rauros into a foaming pit; a glimmering rainbow played upon the fume. And Ethir Anduin he saw, the mighty delta of the River, and myriads of sea-birds whirling like a white dust in the sun, and beneath them a green and silver sea, rippling in endless lines. But everywhere he looked he saw the signs of war.
> (Tolkien, *The Fellowship of the Ring*, Book 2, Ch X, pp. 416–417)

Whatever the device or catalyst—a magical item, a potion, a spell, a place of vision—this trope has become a useful one for providing a sweeping, God's-eye view of the storyworld (made tangible in opening sequences such as the one for HBO's *Game of Thrones*, which emphasises the complex, clockwork-like plotting and mechanistic gamification of Martin's diegetic theatre of war, as brought to life by showrunners David Benioff and D.B. Weiss). Most of all, there is a sense of immanence, of metaphoric significance, in the symbolism that we behold through the eyes of the viewpoint character.

A largely overlooked 'taproot text' of Early Fantasy, *Piers Plowman*, provides textual evidence of the effect of landscape on language, and its terroir upon the imagination. Drabble points out, [t]he strength of [the poem] does not lie in its structure or argument . . .' but in the 'sublimely imaginative poetry', poetic flights of fancy that 'have a sublimity beyond the reach of any other medieval English writer' (1985: 765). *Piers Plowman*, as one of the most popular printed books of the Middle Ages, one of the few books that a wide cross section of society possessed, cast a long shadow. Writers have been wont to nature-inspired reveries, 'slombred in a slepyng', and 'dreaming a dream full of marvels', ever since.

Yet some see this as not always salubrious, or good for the soul. If we are not careful, we could be like Frodo, and be cursed with omniscience.

In *The Anatomy of Melancholy* (1621), Robert Burton warned of 'castles in the ayre', which spending time in natural locations can inspire. Applying the proto-scientific method of his day, Burton provides an exhaustive taxonomy of melancholy, its symptoms, and causes. In a chapter on 'the force of Imagination', a faculty which is 'eminent in all', Burton argues that it has a tendency to 'rageth' in those of a melancholic disposition, making them linger over and amplify whatever they fixate upon. Typical of the scepticism of supernaturalism that followed the Reformation and the anti-Catholic sentiments it encouraged in Protestant countries, Burton saw 'fantasy' as a minor faculty that should be kept in its place: 'this fantasy of ours be a subordinate faculty to reason, and should be ruled by it.' However, in melancholics, 'sleepers', somnambulists, and 'painters, mechanicians, mathematicians' (strange bedfellows, but linked by an ability to visualise what is not present and to make the abstract concrete) it runs wild: 'how many chimeras, antics, golden mountains and castles in the air do they build unto themselves?' He also acknowledges that fear can be the generator of fancy too: 'what will not a fearful man conceive in the dark? What strange forms of bugbears, devils, witches, goblins?'

Despite Burton's advice about not indulging the 'fantastical meditations' such idle reverie can lead to, it has not stopped poets, playwrights, novelists, filmmakers, musicians, and artists from seeking solace and inspiration in nature for centuries.

Some of the greatest works of Fantasy have been inspired by authors playing and roaming in nature as children, and returning for reverie and respite as adults. And these cherished, visceral muscle memories have often been imported into Fantasy storyworlds—whether in this world, or another. A recurring association with natural spaces has provided the mulch of many diegetic universes, as this book will explore. The daydreaming which occurs in such places—by all co-participants of the genre's ecosystem—is at the very heart of Fantasy.

Indeed, such 'fantastical meditations' could be thought of as the very river from which some of the world's greatest writers have been spawned like salmon—a primal, embodied experience that they can spend their whole life striving to return to.

Primal Woods

The playwright and poet William Shakespeare (whose oeuvre qualifies him as an early writer of Clutian 'taproot text' Fantasy: chiefly *Macbeth*, *A Midsummer Night's Dream*, and *The Tempest*) grew up in the predominantly rural Warwickshire, straying sometimes farther than he should. He is thought to have left Stratford-upon-Avon after getting caught poaching at a nearby estate. He drew upon his childhood mythscape in many of his plays. Forests abound in his surviving plays (Barton, 2017)—the Royal Forest of Windsor, a 'woodland near Athens', Birnam Wood—but the most autobiographical was the Forest of Arden, which features prominently in the comedy *As You Like It*. Shakespeare's Arden shares the same name as the once large forest that covered much of the West Midlands: Warwickshire, and parts of Shropshire, Staffordshire, and Worcestershire. Now, somewhat diminished, 'Arden' refers to a district of Warwickshire—a literary toponym that increases property prices. By the time Shakespeare deployed it, it was already becoming a fictional space, hybridised with the portrayal of the Ardennes Forest in Thomas Lodge's earlier romance, *Rosalynde*. In *As You Like It*, the banished Duke, sequestered in umbrageous exile, consoles his party:

> Now, my co-mates and brothers in exile,
> Hath not old custom made this life more sweet
> Than that of painted pomp? Are not these woods
> More free from peril than the envious court?
> Here feel we not the penalty of Adam,
> The seasons' difference; as the icy fang
> And churlish chiding of the winter's wind,
> Which when it bites and blows upon my body,

Even till I shrink with cold, I smile and say
'This is no flattery; these are counsellors
That feelingly persuade me what I am.'
Sweet are the uses of adversity,
Which, like the toad, ugly and venomous,
Wears yet a precious jewel in his head;
And this our life, exempt from public haunt,
Finds tongues in trees, books in the running brooks,
Sermons in stones, and good in everything.
I would not change it.'

(Act II, Scene 1)

An outlaw on the run with a company of players (which is one possible explanation for The Bard's missing years), one can imagine Shakespeare finding a similar consoling fiction in his rustication. As a writer, he has not been alone in finding 'tongues in trees, books in the running brooks', as *this* current work will explore.

Yet, what precisely does the Duke imply by this? It seems to allude to a form of animism—a sense of immanence in which the Earth has a sacralised sentience, and can be communicated with through ritual speech acts, gestures, and offerings. For many indigenous cultures this is a central tenet; and the latest research into neo-animism and biosemiotics seems to confirm this (we shall return to and elaborate upon these concepts), for 'animism never dies', as Gilbert points out (2022: 397), and mythic aquifers have a tendency to rise to the surface again after years of disenchanted, ontological drought.

But to focus on one of the Duke's phrases, 'books in the running brooks'—could this be an allusion to one of Shakespeare's own sources of inspiration? He lived in Stratford-upon-Avon, a handsome market town defined and dominated by the river Avon, which is fed by many tributaries. As a boy, Shakespeare would have played along, on, and in these. And as a budding young wordsmith, their pink noise would have been conducive to composition.

Research in neuropsychology has shown how the sound of running water—by springs, riverbanks, lakes, and the sea—can affect brainwaves, changing them from alpha to theta wave patterns, allowing for greater synaptic connections, which are conducive to 'leaps' of inspiration across the bihemispheric brain (Iain McGilchrist, 2009). Archimedes experienced his famous 'Eureka!' moment in the bath, and possibly not just from the displacement of water caused by his body, but by the very sound of water itself—that most primal of acoustics, evocative of the amniotic fluid of the womb. How many of us find ideas and sounds start to flow in the shower while our craniums are being pommelled by water? Or experience an expansive quality when on the coast, gazing out across the wide horizon of the sea?

Returning to the Duke's speech, 'Tongues in trees' seems to harken back to the groves of the Iron Age 'oak-priests', the Druids, and foreshadows research in Japan into Shinrin-yoku, 'forest bathing' (Qing Li, 2022), which has demonstrated the benefits of spending time amongst trees in terms of mental health and wellbeing. This is apparently the result of photochemicals, and the phenomenological effect upon the body, but these are open secrets that every child who plays in nature knows intuitively. And every writer, accordingly, remembers—investing their depiction of such spaces with childhood enchantment.

Again, enchantment and its antonym, disenchantment, are key concepts here, which we will be returning to throughout this work. But, for now, let us delve into the matter at hand by considering a key text, which stands like a border oak, delineating the heartwood: that zone of primal creativity. As former medical zoologist Robert Holdstock dramatised in his Mythago Wood Cycle (1984–2009), the natural world is conducive to the creative process. The first novel to be published in the series, *Mythago Wood* (1984), starts with the disappearance of the scholar-father, George Huxley, which leads to his two sons (Christian, and his younger brother, Steven) returning in 1947 to the family home, a lodge on the edge of Ryhope: a small section of ancient woodland situated in Herefordshire, on the Borders between England and Wales. They are shocked to discover the woodland's boundaries seem to have *moved*, and it has overwhelmed the lodge. The father's journal is unearthed, and it relates an increasing obsession with the woodland and various 'mythagos', the core concept of Holdstock's cycle.

The author provides several definitions of this slippery term. The first is offered in *Mythago Wood* by Christian Huxley to his brother, Steven—explaining how in ancient woodland the 'aura' around all living things creates 'a sort of creative field that can interact with our consciousness'. Paraphrasing their father, Christian says,

> And it's in the unconscious that we carry what he calls the pre-mythago—that's unconscious that we carry what he calls the pre-mythago—that's *myth imago,* the image of the idealized form of a myth creature. The image takes on substance in a natural environment, solid flesh, blood, clothing, and—as you saw—weaponry.
>
> (1984: 49)

The sons venture into the various 'zones' (ash, oak, thorn, et cetera) of Ryhope asynchronously, resulting in dramatically different consequences.

They find themselves sucked into the mythic landscape of Ryhope, which is an extended portal or 'Time Abyss' (Clute and Grant, 1999: 946–947): the further in you go, the bigger (or older) it gets.

Elsewhere, I have posited that Holdstock's *Mythago Wood* cycle offers an extended metaphor for the creative process (Manwaring, 2018)—the novels are fecund with stories *about* story, how they are created, told, and remembered. Metafictional layers of the story forest are prime examples of works of Fantasy that have not only drawn heavily upon the natural world (as well as Holdstock's scientific knowledge) to create a rich imaginarium but also deeply affected how *this* author experiences woodland. They dramatised mythopoetically what happens when we enter a woodland and experience the Ecological Imaginary (which I shall expand on below). And they have had, for me at least, real-world impact—inspiring not only forays into the forest but also my writing.

The Mythago process articulates the apparently mysterious, often intuitive processes of creativity. As Shakespeare's Theseus says in *A Midsummer Night's Dream*: 'The poet's eye, in fine frenzy rolling, / Doth glance from heaven to earth, from earth to heaven; / And as imagination bodies forth / The forms of things unknown, the poet's pen / Turns them to shapes and gives to airy nothing / A local habitation and a name.' This 'local habitation' could be called Art, a house with many rooms. And as Emily Dickinson wrote, 'Nature is a Haunted House—but Art—a House that tries to be haunted' (Letter to Higginson, 1876: L 459a). It is worth reflecting on this endeavour. It could be argued that the mode of Fantasy is perhaps the most effective 'haunted house' in fiction—one especially suited to the evocation and (re)enchantment of the natural world.

Enchantment, Disenchantment, Re-Enchantment

Many works of Fantasy offer a conscious rebuff to Weberian *Die Entzauberung der Welt* (1917), the closest translation of which is 'the unmagicking of the world'. Weber argued that as a result of the Enlightenment, the hegemony of scientific empiricism and Capitalist materialism, a demystification of the world had occurred by the 20th century, although Matthew Arnold had lamented the 'long withdrawing roar' of the Sea of Faith in 'Dover Beach' (1867), and before that the Romantics had offered the first counterwave to the foreclosure of the imaginative possibilities of the supernatural and spiritual. It is perhaps no coincidence that Weber first lectured on his notion of 'disenchantment' (as it has most commonly become known as) in the final year of the First World War, that vast global tragedy that had shattered irrevocably the Grand Narratives of the Victorian Age and given birth to Modernism. In the aftermath of the atrocity of the 9/11 terrorist attacks in New York and beyond, the concept of *re*-enchantment started to gain traction—first in academic circles, and then in popular culture as the traumatised public sought Tolkienian Recovery, Escape, [and] Consolation (1997: 145–154) in the consoling fictions of Fantasy (e.g. the massive successes of *The Lord of the Rings* trilogy, and the *Harry Potter* franchise). Writers such as Bennett (2001), Landy and Saler (2009), Elkins and Morgan (2009), Graham (2007), and several others have articulated and advocated for this 're-enchantment'.

Yet Crawford has critiqued this counterwave, seeing something provisional and inauthentic in such claims, stating, 'the literature on enchantment seems actually to be uncomfortable talking about enchantment' (2020). Despite his statement that there is 'no literature on re-enchantment' (2020), that is, work that actively encourages a genuine re-enchantment of the world, I would argue that there is a whole body of work that delights in imagining its full-bloodied reality: Fantasy. And more than just positing a simplistic thesis that 'magic exists', many of these texts actively engage with the Weberian axis of disenchantment. There is often an absence, thinning, or seepage from one side to the other—the magical 'other' reality: the more mundane apparent one. A classic example of this is Hope Mirrlees' *Lud-in-the-Mist* (1923). Of course, there are numerous 'crosshatches' (Clute/Grant, 1999) where the ontological divide between a magical reality and a non-magical reality is less clear-cut. Tales where magic has been banned (*Harry Potter*), is forgotten (*The Buried Giant*), or is returning (*Song of Ice and Fire*). The re-enchantment required is part of the healing of the world. It was there in Gilgamesh (where the Campbellian Hero's Journey failed), *Beowulf*, *Sir Gawain and the Green Knight*, *Kalevala*, *Y Mabinogi*, and many others. And it lingers, as a trope, in countless classic and contemporary Fantasy texts (*LOTR*, *The Books of Earthsea*, *The Dark Is Rising*, etc.). And enchantment is not necessarily a static concept, as Zoe Gilbert emphasises in her riotous celebration of the lives and adventures of the British folkloric figure, Herne the Hunter, *Mischief Acts:* 'Enchantment is not in and of itself old-fashioned. It may change its form. It can, like all eternal concepts, respond to the zeitgeist' (2022: 54). Gilbert embodies enchantment in the feral, anarchic figure of Herne, running roughshod over boundaries of time, form, and genre (poetry, folk tale, fantasy, Folk Realism, exegesis). This feels like a far more resilient and fluid personification of the concept—closer to its beating heart.

As the modern world attests, history is definitely *not* static—ideas consigned to its dustbins re-emerge. Similarly, the dis/re/enchantment vortex could be seen as a continuous cycle, an existential Coriolis. Christian poet and scholar of Fantasy Malcolm Guite would go further, refuting the idea of a mechanistic 'clockwork universe', proposing instead (citing Tolkien's cosmology in *The Silmarillion*) that it is a poem—and not a fading echo, but an ever-renewing song (Guite, 2025).

Thus, rather than the negative reading of 're-enchantment' by Crawford as a kind of self-willed illusion, my reading of it suggests a re-connection to that which is vital, actual, and embodied—the innate wonder of the world. Rather than an ontological slippage or degeneration, in which we fail to perceive the actuality of existence, and the reality of the historical moment, I argue that Fantasy instantiates the world, refreshes our awareness of our environment, and reminds us of the infinite potentiality of language and being. It encourages neuro-plasticity, and healthily ludic notions of identity. Not just gender-fluidity, but genre-fluidity. As self-aware genred-beings, we can consciously step in and out of different frames, choosing (or not) to engage with the respective codes we encounter. Thus, in refutation of Crawford's claims that '[t]he "alternative" worlds of genre fiction . . . might offer some solace to the citizens of a disenchanted world, but they don't change the condition of that world' (Guite, 2025), my central thesis here is that Fantasy can do just that.

The stages that have led me to that conclusion have been informed by a lifetime of reading and writing Fantasy, but started formally with a Research Fellowship in 2022, which enabled me to undertake extensive field research at key Fantasy sites across the United Kingdom and the Republic of Ireland. I set out with these questions: How does the natural world inform the writing of Fantasy? And does the reading of Fantasy enrich our appreciation of it? Can the imaginary worlds oft Fantasy have a real-world impact? I also used a research questionnaire to solicit answers from a cross section of writers, readers, and scholars of Fantasy (see Chapter 7: **Fantasy Sites**).

Before picking through any gleanings that have arisen from this research, let us define some terms and parameters.

(Re)Defining Fantasy

To define the parameters of this book, it explores environmental aspects of Fantasy in any form—short story, novel, screenplay, computer game, roleplaying game, graphic novel, LARP scenario, etc. Others have laboured over the definition of Fantasy (Todorov, 1975; Elgin, 1985; Attebery, 1992; Clute and Grant, 1999, Jackson, 1998; Hume, 1984; Mendlesohn, 2008; Sangster, 2023). All of these have had something useful to say about Fantasy, but I prefer to go to earlier and alternative sources in an attempt to find fresh perspectives. Thinking about any subject can become fossilised, and it is easy to become tramlined into thinking about a subject in a certain way because of existing scholarship. Influential critics can dominate the way we perceive something, to the point that it becomes hard to apprehend it except through their cultural filters. I think Fantasy should be less defined by its tropes than by the *effect* it has upon the reader/viewer/listener/player:

an ontological expansiveness or destabilisation, and, apposite to our purposes here, a re-enchantment of the world. These qualities are articulated by those I cite below.

Joseph Addison was to provide a 'working definition' of the Fantastic for the long 18th century in what he called his 'fairy way of writing' (1712):

> Thus we see how many ways Poetry addresses it self to the Imagination, as it has not only the whole Circle of Nature for its Province, but makes new Worlds of its own, shews us Persons that are not to be found in Being, and represents even the Faculties of the Soul, with her several Virtues and Vices, in a sensible Shape and Character
>
> (*The Spectator*, 1712)

This ideation of proto-Fantasy was forged in the Early Modern Period at the fault-line between religious belief and positivist scientific discourse. In a contemporary exegesis on the origins of the Fantastic, Sandner observes, 'The fantastic is an acutely modern literature always coming after primary belief in the supernatural has been emptied of meaning, marking the uncanny return of exploded supernatural belief' (Sandner, 2011: 12). This underlying tension is best articulated in what could be called a High Mimetic style, in which something immanent seeks to break through the plain of materialism. Examples of this effect can be seen in the Gothic, the ghost story, and countless stories of the Weird. Addison goes beyond the 'inherent values' of 18th-century personification in his conception of the Fairy way of writing as not restricted to 'mere allegory' (and in this sense his argument is proto-Tolkienian), and imagines the function of such writing in making 'new Worlds of its own' and other ways of Being. So, more radical than his reductionist commentators (e.g. Wasserman, 1950: 443) would claim, Addison's project has the flavour of the Reverend Robert Kirk (192) about it (Manwaring, 2018, 2021), an air of potentiality, though more formally cloaked in the discourse of the period: 'an inquiry into the psychology of the imagination and a scientific discrimination of the functions and limits of the different arts' (Routh, 1913). It dovetails with the opus of a contemporary of Kirk's, John Aubrey, gentleman antiquarian, whose near-religious fervour in pursuit of scientific enquiry mirrors Kirk's adoption of pseudo-scientific, empiricist methodology in pursuit of religious enquiry (Scurr, 2015: 430). In her experimental biography of the antiquarian, *John Aubrey: My Own Life*, Ruth Scurr comments on Aubrey's modus operandi: 'Aubrey's approach to his own and other lives was imaginative and empirical in equal measure' (2015: 13), which she then adopts in her bold approach, where she creates a fictional diary based upon real historical events. This methodology suggests a useful quality in good Fantasy, in which actuality is the grit in the oyster of the imaginary, and vice versa.

This 'imaginative and empirical' approach is articulated by Hawthorne in his introduction to *The Scarlet Letter*, 'The Custom House' (1850). Describing the harbour-side dwelling in Salem, where, late one evening as he sat alone by the merry hearth he observed pale moonlight falling upon the carpet, 'a medium the most suitable for a romance–writer to get acquainted with his illusive guests'. The mundane becomes enchanted by the magical, and in this liminal space inspiration is nurtured and the Fantastic finds a home: 'Thus, therefore, the floor of our familiar room has become a neutral territory, somewhere between the real world and fairy-land, where the Actual and the Imaginary may meet, and each imbue itself with the nature of the other.' (1850)

This cross-fertilisation is at the heart of the Ecological Imaginary. The Primary World inspires writers, who create worlds inflected with its actuality (albeit in a tilted or transmogrified form). These in turn influence readers, listeners, viewers, and players who see the Primary World anew (either through its association with a Fantasy text or via the palimpsest of a transformed consciousness in which layers of the Imaginary are transposed upon the everyday—as a lingering after-effect, or deliberate make-believe).

Hawthorne is not the only author to find fertile ground in the creative tension between opposites. In a letter to his younger brothers, George and Thomas (21 December 1817), the Romantic poet John Keats outlined his concept of 'negative capability': when a man is capable of being in uncertainties, mysteries, doubts, without any irritable reaching after fact and reason'. In the same poem, Keats seems to allude to Coleridge's 'fragment of a dream', the poem 'Kubla Khan', the completion of which, inspired by an opium-induced reverie, was famously interrupted by a 'Man from Porlock' on business. Keats seems to suggest that its very efficacy is in its incompleteness—in a refusal of closure.

From Coleridgean 'suspension of disbelief' to Keatsian 'negative capability', this self-willed credulity and anti-omniscience is the very antithesis of the Enlightenment project. The rejection of supernaturalism that began with the Reformation and became enshrined in the Age of Reason left a telling absence in culture and consciousness, which became slowly filled with the products of the imagination, eventually giving birth to Romanticism. As Goya warned in his engraving, 'El sueño de la razon produce monstruos', the 43rd panel from his 1799 album of prints, *Los Caprichos* ('The Fantasies'), 'The Sleep of Reason Produces Monsters.' The engraving depicts a writer slumped over a desk, exhausted. Behind him a whole unruly parliament of owls and bats flap, flit, hoot, and shriek around him, reifying his terrors. Yet, that which is alluded to is so much powerful than that which is literalised—a deferment or sublimation. Goya possibly suggested this in one version of *Los Caprichos,* which contains captions. Beneath Plate 43 it reads, 'Imagination abandoned by reason produces impossible monsters: united with her, she is the mother of all arts and the source of their wonders.'

So, a middle way is suggested, where the inherent tensions are not exploded, but contained. It is this 'bonny road' between extremes where lies Faerie, as Thomas the Rhymer found (Manwaring, 2024), and it is to this territory we shall explore further.

Keeping the Door of the Id Ajar

Keatsian uncertainty stuck a jam in the door of the Id, which later M. John Harrison called 'the transliminal moment' (cited in Mendlesohn, 2008: xxiii)—a lingering on the threshold, or a deferment of ingress, preventing closure, the full reveal. It is akin to the affect of the Sublime, but, to drill down into the detail, it is in nuanced interstices of Tolkienian consolation (Tolkien, 1997: 109–161) and more negative readings of the Sublime as being about 'blockage' or 'usurpation' (Sandner, 2016): 'The tension between these two different outcomes is the dynamic energy of the fantastic itself for readers' (2011: 27). It is as though the battery of the Fantastic is charged by the alternating currents of Mythos and Logos—by a mythic awareness, and a rationalist sensibility. Neither one should achieve dominance indefinitely. And, indeed, many stories oscillate between the two poles—the dramatic charge generated by the friction between them. These are often personified as characters, such as Shakespeare's Caliban and Prospero, or by what John Fowles calls 'The Wild Man and the Academic'.

These qualities are there, in some form, within every writer. For Robert Louis Stevenson, the inchoate autochthonic forces of creative generation he characterised as 'brownies' (1888). Stevenson, recounting examples of particularly vivid dreams—'that small theatre of the brain' (1888: 122)—posits the nocturnal machinations of presences ('players') which bequeath him inspiration. He identifies them as Brownies, benign beings usually associated with households. If treated kindly (e.g. left a saucer of cream), they enact useful household chores, cleaning, polishing, preparing, and so forth. The Reverend Robert Kirk mentions them in *The Secret Commonwealth of Elves, Fauns and Fairies*: 'The Brounies, who in some families are Drudges, clean the house & dishes after all go to bed, taking with him his portion of food, & removing befor [sic] day break' (2006 [1692]). Are these 'Drudges' the genius loci of dream states? Super-terrestrial beings who access human consciousness through the backdoor of the brain? The foot-soldiers of Fancy? Whatever their nature and providence, it seems an appropriate metaphor for the actions of the un- or subconscious, which prior to Freud and

Jung, Stevenson would not have had conceptual framework for. Bound by his literary milieu and available metaphors, he charmingly describes these 'Little People' as a company of unruly players awaiting his authorial direction: 'the little people who manage man's internal theatre had not as yet received a very rigorous training; and played upon their stage like children who should have slipped into the house and found it empty' (1888: 124). And yet even without his intervention, it would seem this lively troupe of 'rude mechanicals' soon take matters in hand, even despite his conscious implementation of craft: 'And yet how often have these sleepless Brownies done him honest service, as he sat idly taking his pleasure in the boxes, *better tales than he could fathom himself*' (1888: 125; my emphasis). It would appear that the Brownies have their own agency, which suggests an autonomous race dwelling within (or accessed through) his mind: 'My Brownies . . . who do one-half my work for me while I am fast asleep, and in all human likelihood, do the rest for me as well, when I am wide awake and fondly suppose I do it myself' (1888: 126). This is an argument foreshadowed by Kirk in the late 17th century, who described them as 'Hollanders or Hollow-cavern inhabitants [who] live and traffick among us, in another State of Being without our Knowledge' (2006 [1692]), and further developed by the Irish mystic, AE, a.k.a. George William Russell in *The Candle of Vision*: 'at times in meditation there broke in on me an almost intolerable lustre of light, pure and shining faces, dazzling processions of figures, most ancient, ancient places and peoples, and landscapes lovely as the lost Eden. These appeared at first to have no more relation to myself than images from a street without one sees reflected in a glass; but at times meditation prolonged itself into spheres which were radiant with actuality' (1918: 29). It is a tantalising notion, and aligns with my hypothesis (2018) that many examples of the Fairy Tradition (the entering of hollow hills, the receiving of artistic gifts, the fixation upon an unattainable object of desire, a certain feyness) articulate the creative process in a metaphoric, or figurative way—limning the secret agency of the consciousness.

This is not to fillet out the supernatural aspect and claim all Fairy folklore can be explained away. Such apparently divergent paradigms may not be mutually exclusive, but part of the same phenomenon, or certainly on a continuum. Stevenson relates examples of vivid dreams at length, as illustrations of this detailed, indigenous creation of narrative material, but also as a pathology which leads to the reification of such dream residuum in the *Strange Case of Dr Jekyll and Mr Hyde (1886)*—a novel that seems to be an attempt to create a visceral Mystery Play of this double-life: the waking man unable to control the wild nocturnal adventures of his dreaming self. Much has been written on Stevenson's Gothic masterpiece—that it is a deconstruction of a hypocritical society, an exploration of the Id, a dramatisation of the demi-monde, and so forth, but as Stevenson was prone to taking laudanum, it seems just as likely to be the 'raw-head and bloody-bones' of his own nightmares. He is in good company with fellow Lotophagi such as Coleridge, De Quincey, and Poe. Aldous Huxley was aware of Stevenson's essay, and it is an interesting footnote that the English intellectual experimented with hallucinogens to open his own 'doors of perception', perhaps keen to awaken his own inner 'Brownies'. Can they be artificially invoked? Or are they endogenous to a writer's genius? This question has haunted those interested in creative writing for centuries. Some would give no credence to the (Classical and Romantic) notion of inspiration, and yet even the sceptics appreciate it when it strikes. Writers crave the adrenaline rush of inspiration and seek it wherever they can, but it often comes unbidden. One of the best descriptions of it is in Michael Chabon's *Wonder Boys*:

> The midnight disease is a kind of emotional insomnia; at ever conscious moment its victim—even if he or she writes at dawn, or in the middle of the afternoon—feels like a person lying in a sweltering bedroom, with the window thrown open, looking up at a sky filled with stars and airplanes, listening to the narrative of a rattling blind, an ambulance, a fly trapped in a Coke bottle, while all around him the neighbours soundly sleep.
>
> (2008: 20)

Chabon's evocative term, the 'midnight disease', redolent of both guilty pleasures and parasomniac activity was seized upon by neurologist Alice W. Flaherty in her examination of the pathology of writerly compulsion and method (2005). She cites examples of hypergraphia, the overwhelming compulsion to write—what could be called Obsessive Narrative Disorder—which allowed the writer no more control than an addict over his vice. But it is a form of mania from which emerge works of great depth and daring, and Stevenson would have been familiar with it: there can be no Hyde without Jekyll.

Although Iain McGilchrist (2009) forensically deconstructs the crude distinction of left- and right-brain activity (the former being 'literal and logical', the latter 'lateral and artistic'), in his examination of hemispherical activity, could the inchoate, minute, and infinitely complex operations of the mind be ultimately what Stevenson was groping towards, using an arresting (and relatable) analogy? Stevenson never asks us to believe literally in his Brownies, just to believe that something indubitably happens beyond his volition to enable him to write and to write with success: 'I am awake, and I live by this business; and yet I could not outdo—could not even equal—that crafty artifice' (1888: 126).

The mysterious processes of Stevenson's 'crafty artifice' bring us close to the heart of Fantasy, even when it was common for his contemporaries to offer some kind of apologia, as Henry James explained: 'It is still expected, though perhaps people are ashamed to say it, that a production which is after all only a "make believe" (for what else is a "story"?) shall be in some degree apologetic—shall renounce the pretension of attempting really to compete with life' (1884 [1888]). What was once thought the gift of the Muses is relegated by the Late Victorian period to mere 'make-believe.' And yet even in an age when the colonising empiricism of the Empire sought to taxonomize, categorise, collect, and appropriate the world's riches, some authors sought to chart the Ultima Thule of the imagination. Edgar Allen Poe, in his essay, 'The Importance of the Single Effect in a Prose Tale' (1842), discussing Hawthorne's 'Twice-Told Tales', summarised the qualities most effective in a short story—chiefly a unity of effect. He was to elaborate upon this in his essay, 'The Philosophy of Composition' (1846), which focused on his poetry, chiefly the composition of his Gothic poem, 'The Raven'. In discussing the evolution of his Weird fiction (for *Fiction Writers on Fiction Writing*, 1923), Algernon Blackwood describes how his chief source of inspiration is nature itself: 'The genesis of a story with me is invariably—an emotion, caused in my particular case by something in nature rather than in human nature: a scrap of colour in the sky, a flower, a sound of wind or water; briefly, an emotion produced by beauty' (quoted in Hoffman, 1923: 12). These ideas he expanded upon in a later essay 'The Genesis of Ideas' (1937) republished with an introduction by Henry Bartholomew as *The Unknown: weird writings, 1900–1937* (2023). An idea for a story is triggered, Blackwood claims, by a 'strong, deep emotion that seeks, naturally, to express itself'. This emotion 'may be caused, let us say, by beauty of nature, rapture of music, wonder, horror of the supernatural, but, whatever the cause, the emotion, boiling and bursting for expression, seeks an outlet'. There the emotion remains in 'that enormous storehouse of things unused, unexpressed—the subconscious,' until 'some chance click of the cupboard latch releases it' (2023: 194).

In the early 20th century, M.R. James emphasises the need for verisimilitude and tonal restraint 'reticence', arguing that the greatest effects are created by a refusal, deferral, or redaction of the supernatural: 'The reading of many ghost stories has shown me that the greatest successes have been scored by the authors who can make us envisage a definite time and place, and give us plenty of clear-cut and matter-of-fact detail, but who, when the climax is reached, allow us to be just a little in the dark as to the working of their machinery' (1929).

Although James warns that, in fiction at least, 'We do not want to see the bones of their theory about the supernatural' (1929), it is essential for students and scholars of Fantasy, in its broadest sense, to boil up the bones. Hopefully, this present broth will be suitably appetising, or at least useful in some way.

A Phenomenological Definition of Fantasy

Addison, Hawthorne, Stevenson, Poe, and James all help start to limn a phenomenological definition of Fantasy, which could be summarised as one that foregrounds the experiential. The lifeworld of the author directly or indirectly inflects the formation of a Fantasy text, which in turn influences the reader—both in the act of reading and in its impact upon their experience of their own lifeworld. Reality is made more plastic, and ontologies shift. Awareness of the overlapping lifeworlds we share this biosphere with is enhanced. Our senses may be heightened; our actions influenced. As a result, our very beingness in the world may be fundamentally altered. The inchoate qualities of our identity/identities may feel validated. The world is suddenly more polysemous, and so are we. This is, of course, a 'best case scenario', and a Fantasy text can be simply enjoyed as a comfort read, view, or listen. We can enjoy a Fantasy game, cosplay, or LARP session, without it being life-changing. And yet all of these foreground the *experience* of Fantasy, which is rarely discussed, but is for many of us the sine qua non of why we love it. A phenomenological definition foregrounds this and the genre's multivalent affordances.

This does not refute other attempts to define Fantasy, which is, by its very nature, a broad church, or deliberate 'fuzzy set' (Mendlesohn, 2008). Many attempt to pin it down, to stake their claim in this *Terra Incognita,* yet none could truly claim to fully capture its extent. Its edges fade into the mist the closer we get to them, for its territories are continually shifting and expanding. Any attempt at canonicity or authoritative delineation is doomed to fail, even if well-informed traverses of its vast terrain are worthy endeavours. It echoes the manner in which Sandner locates 'the fantastic within the locus of competing claims' (2016: 32). In the space-between of this locus, the best work, I argue, emerges.

Building upon this broader definition of the mode, I now wish to clarify what I mean by the 'Ecological Imaginary'.

The Ecological Imaginary

Although the Ecological Imaginary has deep roots through human (and more-than-human) history, it was first coined as a term by Matthew Gandy (2018) within an architectural context as a biomimetic reframing. In describing its origins within his discipline, Gandy explains that, 'ideas drawn from nature have played a significant role in developing an "ecological imaginary" in which ideas or metaphors drawn from the biophysical and medical sciences have been used to understand the form and function of the modern city' (Gandy, 2018: 55). Beyond this 'implicit naturalisation of urban processes', Gandy seeks to expand the definition from 'organicist conceptions of the urban form' to one that challenges such tropes (which can easily become so much greenwash), offering an 'alternative set of readings', which foreground 'malleable, indeterminate and historically specific dimensions' (Gandy, 2018). From Gandy's ideation we can take not only the term (which I redefine below within the context of Fantasy) but also his desire to look beyond the aesthetics to methodological and theoretical elements. As we shall discover below, it is not enough for a Fantasy text to feature trees, mountains, rivers, or wildlife. This is not sufficient to make it ecological, and it is not a true exploration of the Ecological Imaginary.

Before we attempt a definition, let us look at a couple of sister concepts that may help us conceptually parse our version. Firstly, there is the inversion, 'imaginative ecologies', posited by Villaenueva-Romero, Kerslake, and Flys-Junquera (2021). They suggest that by 'using imagination as a tool' (2021: 2), we can use the Arts and Humanities to explore environmental issues—through the production of cultural artefacts and the discourses they generate. Expanding upon the idea that 'artistic works are themselves like cultural ecosystems' (2021: 2), through the way they can forge connections, generate debate, and raise awareness, the editors define 'imaginative ecologies' as 'ecologies of the mind moved by the affective connection between the writer/artist and the natureculture s/he is engaged with during the creative process of acknowledging the call of the more-than-human in an anthropocentric world where boundaries are built between "us" and the "other"' (2021: 2). Leaving the unwieldiness of this definition aside, the editors locate their imaginative ecologies within the (human) mind, even as they emphasise the 'call of the more-than-human', as though it is a purely conceptual process.

In my notion of the Ecological Imaginary, I argue that the process is far more visceral, and phenomenological than that.

Even though the editors above *do* acknowledge the need for literary and artistic work to be embodied and situated if it is to generate change, their focus is biased towards the Environmental Humanities more than the Arts, and is too wide for our genre-specific purposes here. Also, I argue that the creative processes and unique affordances of Fantasy are *especially* suited to the project, because it has evolved within an environmental awareness that has inflected it since its earliest iterations: that Fantasy has grown from this 'tree'—the Tree of Life itself—and not just been grafted onto it.

Perhaps closer to our purposes here is Westling's notion of the 'Eco-Literary Imaginary' (2016), which suggests human cultural production—from the earliest cave paintings to the blockbusters of the present day—all carry the 'silt' of Merleau-Ponty's 'imaginary', the sine qua non of existence as embodied beings immersed within the 'flesh' of the world, the zone of interpenetration between all that is: 'All these works are sedimentations of an Ecological Imaginary nourishing all human cultures' (2016: 65). Paraphrasing Ponty, Westling argues that 'unless we recover this sense of a wider intersubjective and interdependent reality, our world will perish' (2016: 66). Although Westling's focus is very much on the phenomenological, she usefully brings into discussion notions of sedimentation, biosemiotics, and entanglement (all key concepts here). In the latter especially, she is most eloquent, emphasising how the eco-*literary* imaginary is never merely conceptual. It is acted upon and acts upon the world: 'Literary works, like language in general and other cultural creations such as painting and music, are ecologically and intersubjectively involved with the lifeworld their creators experienced, though we may have forgotten how fully they encode ecological memory and allow for adaptations to newer realities in their audiences' (2016: 74). Westling alludes to two key concepts from phenomenology, Husserl's *Lebenswelt*, or Lifeworld (the lived experience of reality specific to each sentient species) and Uexküll's *Umwelt*, 'the lived reality of interaction with the environment for our species, distinct though overlapping with those of other organisms' (2016: 68). The first could be said to be 'singular consciousness', the latter, 'plural, or relational consciousness.' All of these are useful concepts for our discussion here, but risk limiting us to one theoretical or linguistic frame—Merleau-Ponty's 'chiasmic ontology' (2016: 67). And Westling's eco-literary Imaginary again is too broad to be *entirely* useful for our purposes here.

And so triangulating between Gandy's 'ecological imaginary' (2018), Villaenueva-Romero, Kerslake, and Flys-Junquera's 'imaginative ecologies' (2021), and Westling's 'eco-literary imaginary' (2016), I suggest the Ecological Imaginary as follows, acknowledging that any definition is subject to critique, revision, and expansion.

The Ecological Imaginary is a conceptual framework for summarising how Fantasy texts (in the broadest sense, to include media such as film, TV, computer games, graphic novels, audio dramas, roleplaying games, interactive fiction, etc) explore organic lifeworlds, ecological thinking, and environmental issues. In such texts (super/supra), natural worlds are not mere backdrops, but an *intrinsic part* of the plot. Their affect is foregrounded, and may emphasise certain key themes, such as mistreatment of the environment. The more-than-human often has not only sentience but agency. Issues around ecological balance, ecological karma, and/or justice are explored. In texts created since the Rio Earth Summit of 1992, Climate Change (Crisis/Emergency) should ideally be acknowledged in some way, however metaphorically, although it is accepted that *all* cultural artefacts are inflected by this in the 21st century (Bould, 2021). Yet, such texts should avoid the didactic, especially single positions and simple solutions. And finally, the Ecological Imaginary can be a way of exploring natural worlds through the mode of Fantasy—perhaps dramatising a sense of a sacralised nature, immanence, and other neo-animist beliefs. It can seek to imagine, and even inspire, a re-enchantment of nature.

Having provided a working definition, I shall now turn to the wider critical field.

Key Critical Works

Of course, I am not the first to explore this aspect of Fantasy, although earlier critical works on Fantastika rarely discuss environmental aspects beyond considerations around setting. It is perhaps more common in the scholarship of science fiction (e.g. Kim Stanley Robinson's *Green Planets* anthology of essays). Even works such as the promisingly titled *The Nature of Middle Earth* (Hostetter, 2021) focused more on aspects of world-building. Although some ecological themes were foregrounded in scholarship around Robert Holdstock's Mythago Wood (e.g. the 2011 D.E. Morse and K. Matolcsy (eds) *The Mythic Fantasy of Robert Holdstock: critical essays on the fiction.* Jefferson, NC: McFarland, including my own paper for *Writing on Practice* in 2018: '*Ways Through the Wood: The Rogue Cartographies of Robert Holdstock's Mythago Wood Cycle as a Cognitive Map for Creative Process in Fiction*'), it is only relatively recently that these wider connections across the mode of Fantasy as a whole have been made explicit.

The earliest exploration of ecological aspects of Fantasy can be found in Donald Elgin's *The Comedy of Fantasy* (1985). The central thesis that Fantasy is in the comic mode, and that this is inherently ecological, is a limited and unconvincing one. It is also written before anthropogenic Climate Change and the Anthropocene were tangible scientific concepts. I would argue for Fantasy's eco-credentials outside of this narrow frame. Nevertheless, an interesting early foray.

Adopting a rather eccentric approach, *The Ecology of Middle Earth* (Juhren, 1994) offers an ecological audit of Middle-Earth, treating it as though it were a real biome. Although lacking in critical rigour or insight, it does provide a useful survey of the flora and fauna Tolkien populated his Secondary World with, and does, in its conclusion, make an interesting point—that the worlds of Middle-Earth and Lorien (*thinned* and *poldered,* respectively) blend in the Third Age, creating a re-enchanted realm (where The Shire is reforested with trees grown from the seeds gifted to Sam Gamgee by Galadriel). In *Defending Middle Earth* (1997) Patrick Curry offers a single chapter of the nature and ecology of Tolkien's realm. In *Representations of Nature in Middle Earth* (2015) Simonson et al explore the biodiversity of Tolkien's Secondary World through cross sections of close-focus essays. And in *Mapping Middle-Earth* (2024) Behrooz Focuses on the cartographical aspect, an area I explore across the genre in 'Mapping the Fantastic'.

A substantial contribution to the field, but one that does not acknowledge the contribution of oral narrative, include other media, or discuss real-world sites and impact as I do here, is the collection of essays edited by Attebery, Oziewicz, and Dedinová—*Fantasy and Myth in the Anthropocene* (2021). The anthology is an important contribution to the field of scholarship in the Ecological Imaginary, but it is not, as it claims to be, 'The first study to look at the intersection of the discourse of the Anthropocene within the two highly influential storytelling modes of fantasy and myth'. *Storytelling for Nature Connection*—originally *Storytelling for a Greener World* (2014)—consisting of twenty-one chapters from a range of creative-critical practitioners, predates *Fantasy and Myth in the Anthropocene* by seven years. Coming from a smaller press, this has gained less attention; however *Storytelling and Ecology*, by one of its editors (Anthony Nanson), was also published by Bloomsbury in the same year as the anthology by Attebery, Oziewicz, and Dedinová, and offers a rigorous defence of storytelling in its truest form—as oral narrative—as a mode for cultivating ecoliteracy and engagement with the environment and environmental issues.

Since the publication of his seminal work, *Strategies of Fantasy* (1992), Attebery has long been an authority on the genre. In *Fantasy: how it works* (2022) he offers an excellent, insightful overview of the genre, with a thorough analysis of its mechanisms, but it is not a thesis predicated on environmental aspects (it is strongest in its discussion of the political aspects). The exception is his 'mitochondrial theory of Fantasy' (2022: 80–94) which provides a multivalent biomimetic model of the 'ecosystem' of Fantasy, in terms of writers, readers, fans, etc. It suggests that any text exists within a lattice of other texts, a parasitic-mutualistic (Sheldrake, 2020) network that influences without absorbing completely new nodes. 'Thinking in mitochondrial terms reminds us that a text that is drawn into a newer text is still alive, still working, still itself. The metaphor invites us to see how host and symbiote both benefit from the relationship. And it shows that a thing so familiar as to be ignorable—a cell, an allusion—is, in reality, much stranger and more complex than we realize. A simple act of reference is really a whole history of incorporation, negotiation, and synergy' (2022: 90). This idea provides a theoretical framework for what Jonathan Lethem calls, in direct refutation of William Bloom's literary 'anxiety' (1997), 'the ecstasy of influence' (Lethem, 2007). Acknowledging that 'literature has been in a plundered, fragmentary state for a long time', Lethem advocates a rejoicing in this, rather than a denial of it. He cites the 'open source' culture of Blues and Jazz, and Hip Hop; the avant-garde Art Movements of the 20th century; Burroughs' cut-and-paste; the speeches of Martin Luther King; and Bob Dylan's music, whose 'originality and his appropriations are as one'. Reflecting on these, Lethem observes, 'it becomes apparent that appropriation, mimicry, quotation, allusion, and sublimated collaboration consist of a kind of sine qua non of the creative act, cutting across forms and genres in the realm of the cultural production'. Thus, Attebery's theory provides a resilient, biomimetic frame for such processes. How this now looks in the light of Generative AI—the rapidly mutating 'virus' that rapaciously scrapes existing work from the internet and regurgitates it instantaneously in endless variations, depriving authors of royalties in the process—we are yet to see. Mitochondrial influence only works if there is a continually renewing 'gene pool' of creative content; not in a paradigm of generic and sterile 'Terminator-seed' artificial content: a law of diminishing returns.

Stroud's *Historical Dictionary of Fantasy Literature* (2023), an otherwise useful survey of the genre, dedicates less than a dozen lines to the Anthropocene. Many of its definitions lean heavily upon those of Clute and Grant (1999), in whose *Encyclopedia of Fantasy* we find useful terms such as Taproot Texts, Wainscot, Crosshatch, and Thinning.

Sangster's *An Introduction to Fantasy* (2023) provides a comprehensive and insightful overview of the formations and affordances of the genre, but only touches lightly upon environmental aspects. This area is made slightly more explicit in his co-edited volume, *Realms of Imagination* (Kirk and Sangster, 2023), published as part of the major exhibition at the British Library that year. Divided into four sections (Fairy and Folk Tales; Epics and Quests; Weird and Uncanny; Portals and Worlds), the various essays and exhibits touch upon some environmental aspects, in the broadest sense, especially with regard to world-building.

Having surveyed key scholarship (with the proviso that new work is always coming to light), I shall now consider the theoretical fields within which this current investigation into environmental aspects of Fantasy will operate. These primarily consist of ecolinguistics, phenomenology, and biosemiotics, with elements of inclusionality, systems theory, and intersectionality.

An Ecosophic Frame

I think the best frame with which to analyse the ecological credentials of Fantasy is via ecolinguistics (a term first formulated in the early 1990s when the impact of Climate Change broke through into popular consciousness via waves of news stories, and the international platform of the Earth Summit in Rio de Janeiro, 1992). Adopting an 'ecosophic frame' (Stibbe, 2021), we can conduct a deeper analysis of a Fantasy novel: concentricities of analysis textual (micro), thematic (mezo), critical and reader response (macro), and, finally, fan culture (meta). Before we look at those, first let us consider the definition of ecosophy by the godfather of ecolinguistics, Arne Naess:

> By an ecosophy I mean a philosophy of ecological harmony or equilibrium. A philosophy as a kind of sofia (or) wisdom, is openly normative, it contains both norms, rules, postulates, value priority announcements and hypotheses concerning the state of affairs in our universe. Wisdom is policy wisdom, prescription, not only scientific description and prediction. The details of an ecosophy will show many variations due to significant differences concerning not only the 'facts' of pollution, resources, population, etc. but also value priorities.
>
> (cited by Drengson and Inoue, 1995: 8)

Arran Stibbe, in *Ecolinguistics: Language, Ecology and the Stories We Live By* (2021), expands and updates Naess' definition. Stibbe foregrounds a biophilial-instrumentalist approach: 'In ecolinguistic analysis of an ideology, the question is not whether it is true but whether it encourages people to preserve or destroy the ecosystems that support life' (Stibbe, 2021: 22). This touches upon the central tenet of this book—that Fantasy supports this ecolinguistic agenda, even as it dramatises a spectrum of attitudes and behaviours towards living systems and life-forms. Stibbe suggests ecolinguistic discourse expands the collective imaginary: 'The purpose for analysing beneficial discourses (which encourage more ecologically beneficial behaviour) is to promote them as useful alternative ways of telling stories about the world and help them become widespread even if they are currently relatively unknown' (Stibbe, 2021: 26). Fantasy, an inherently metanarrative genre (Sangster, 2023) continually haunted by itself—a biome of narratives and narrators, listeners, readers, and texts—indubitably explores 'alternative ways of telling stories'. Through its massive fan base, it can certainly help ecological thinking reach wider audiences. The languages of Fantasy—that frequently give voice to the more-than-human—help us to expand our sphere of empathy beyond the limits of our own *umwelt*. They suggest we live in a polyphonic universe, one with myriad paradigms, as opposed to the monotone monoculture of the reductively material. Stibbe emphasises, 'It is when beneficial discourses start finding themselves incorporated within the mainstream discourses that surround us that they can start to have an impact on the stories we live by' (2021: 30).

Adopting an ecolinguistic approach in our survey of environmental aspects of Fantasy, one could deploy concentricities of analysis:

1. textual (micro)—a close reading of the Fantasy 'text' (novel, short story, film, computer game, graphic novel, etc).
2. thematic (mezo)—drawing out any commonalities of themes from comparative texts (e.g. depictions of the more-than-human).
3. critical & reader response (macro)—the cultural impact of a text, in terms of fan and professional reviews, discourse on social media, the 'buzz', awards, etc.
4. fan culture (meta)—fan engagement with the storyworld of the Fantasy text, for example, LARPing, cosplay, fanfic, etc.

This changing of frame I call 'tree-ring thinking'. One can expand infinitely outward or inward in any given situation. This can shift entrenched opinions and attitudes, and provide fresh perspectives and insights. The symmetries and affinities that can be discerned in orders of magnitude can also help us identify the patterns that we dance by. It is the fractal leaf that mirrors the tree it has grown from, or the Fibonacci of a nebula.

The Flesh of Fantasy

Overlapping the lifeworld of Naess' ecosophy, we have the *Lebenswelt* of Husserl's phenomenology. To first define the term,

> Phenomenology is the study of structures of consciousness as experienced from the first-person point of view. The central structure of an experience is its intentionality, its being directed toward something, as it is an experience of or about some object. An experience is directed toward an object by virtue of its content or meaning (which represents the object) together with appropriate enabling conditions.
>
> (Woodruff Smith, 2013)

Building on the groundwork of Edmund Husserl, Maurice Merleau-Ponty expanded and refined his ideas. Merleau-Ponty conceived of 'flesh'—the sensory interplay between the perceiver and the perceived, a mutualistic membrane, which seems apposite to our focus here (the cross-fertilisation between language and environment). Language, Merleau-Ponty seems to suggest, has its own 'landscape' or sensorium:

> Language, as a way of describing and referring to the world, informed by sensory phenomena, can not be anything other than such itself—the

> inherent 'fleshiness' of language, made of sounds and shapes that as just as much part of the global ecology as rainforest, polar bear, and sperm whale. (1962)
>
> Language is first and foremost an embodied experience—we understand directly through the senses in a pre-verbal way. Merleau-Ponty summarises this process as meeting, encounter, participation: '. . . meaning sprouts in the very depths of the sensory world, in the heat of meeting, encounter, participation.'
>
> (Abram, 2017 [1997]: 75)

Spoken words have a physical presence, an 'affective tonality' (2017 [1997]: 75). These have an impact upon the listener: 'the soundful experience of spoken words upon the sensing body.' (2017 [1997]: 79)
Fundamentally, language reconnects us to the natural world:

> 'Language as a bodily phenomenon accrues to *all* expressive bodies, not just to the human. Our own speaking, then, does not set us outside of the animate landscape but—whether or not we are aware of it—inscribes us more fully in its chattering, whispering, soundful depths.'
>
> (2017 [1997]: 80)

Merleau-Ponty encapsulates this with his ideation of 'eco-vocal expression', speculating that '. . . if human language arises from the perceptual interplay between the body and the world, then this language "belongs" to the animate landscape as much as it "belongs" to ourselves' ([1997]: 82). The web of language binds us to the web of life: 'the organic, interconnected structure of any language is an extension or echo of the deeply interconnected matrix of sensorial reality itself' ([1997]: 84). Ultimately, Merleau-Ponty, declaims: 'it is not the human body alone but rather the whole of the sensuous world that provides the deep structure of language' ([1997]: 85). This idea that language evolves out of its contextual ecosystems reaches to the core of the Ecological Imaginary.

Signs of Life

The third major theoretical field I would like to draw upon here is the emerging science of biosemiotics: 'Biosemiotics is an interdisciplinary research agenda investigating the myriad forms of communication and signification found in and between living systems. It is thus the study of representation, meaning, sense, and the biological significance of codes and sign processes, from genetic code sequences to intercellular signal[l]ing processes to animal display behavio[u]r to human semiotic artifacts such as language and abstract symbolic thought' (https://www.biosemiotics.org/what-is/). In the conclusion, I would like to posit a Biosemiotic Theory of Fantasy.

The three 'secondary' theoretical fields of influence are inclusionality, systems theory, and intersectionality. That there is no separation in this phenomenological way of experiencing and interpreting/interpenetrating reality is affirmed by the notion of 'inclusionality' posited by mycologist Alan Rayner (2010). Arguing against discreetist thinking by drawing upon an interdisciplinary study of biological systems, Eastern philosophy, painting, and poetry, Rayner advocates 'natural inclusionality' as a challenge to the destructive hegemonies of 'selfish gene' discourse, which began with the Cartesian Divide and culminates in the rapacious endgame of Neoliberalism:

> All that may ultimately be needed to unlock ourselves from this unnatural confinement imposed by abstract rationality is the simple understanding that space cannot be cut, confined or excluded and so is a continuous presence throughout and beyond the energetic boundary interfacings of natural Figures as flow-forms. By the same token, boundaries are energetic interfacings between inner and outer realms, not fixed limits. This simple move from regarding space and boundaries as sources of discontinuity and discrete definition to sources of continuity and dynamic distinction provides the basis of what, in ecological and evolutionary terms has been called 'natural inclusionality'.
>
> (Rayner, 2010)

Rayner's concept of 'dynamic boundaries' is one that serves us well in our multimodal study of Fantasy—the ultimate 'fuzzy set' (Attebery, 1992; Mendlesohn, 2008). This is especially apposite as we focus on environmental aspects of the genre, which are never restricted to only one type (i.e. 'serious' environmental Fantasy that tackles Climate Change head-on, versus 'light' Fantasy that may have tangible natural settings, critters, and scenes, but does not acknowledge the vulnerability and health of such biodiversity). Tropes and discourses have a tendency to leak into each other, *especially* when we try to keep them separate. Rather than resist this 'contamination anxiety' (Lethem, 2007), we can revel in it.

Dancing With Systems

A conceptual framework for how component elements of a system relate both between themselves and with other systems, Systems Theory, was first proposed by biologist Ludwig von Bertalanffy in the 1940s (and later developed by W. Ross Ashby and George Bateson). Bertalanffy suggested the idea of open systems, which interact across systems and avoid the risk of entropy (hardwired into the Second Law of Thermodynamics). This would have been a radical proposal in a period of extreme territoriality, when the world was riven by conflict over borders. Yet philosopher Donella Meadows suggests something even more radical: *dancing* with systems (2017 [2004]). In the light of Quantum Theory, Chaos Theory, Climate Change, and the perpetual entanglements and curveballs of geopolitics and daily life on a planet of billions of interconnected beings (where 'black swan' events are becoming *less* unusual), Meadows urges us to acknowledge this, and even see opportunity within it:

> Let's face it, the universe is messy. It is nonlinear, turbulent, and chaotic. It is dynamic. It spends its time in transient behavio[u]r on its way to somewhere else, not in mathematically neat equilibria. It self organizes and evolves. It creates diversity, not uniformity. That's what makes the world interesting, that's what makes it beautiful, and that's what makes it work.
>
> (2017 [2004])

Accepting this, and not being overwhelmed by it, is the first step towards working *with* such a shifting field of variables—a load of molecules bouncing around within a container; a random cross section of dancers on a dance floor all dancing in their own style; a busy city street; Bartholomew Fair; Goblin Market; Wonderland; Gormenghast; Camelot; Hogwarts; Earthsea; Westeros. The 'special world' where defamiliarisation and cognitive estrangement occurs, whatever Darko Suvin says (although he begrudgingly acknowledges it does when he widens his purview of Fantasy from a very limited and hoary sample). Where old habits and opinions are challenges; new rules must be quickly learned; allies discerned from enemies; the true Grail from the many false.

By learning the 'dance' (following Meadows first steps of 1. Get the beat, 2. Listen to the wisdom of the system, 3. Expose your mental models to the open air, 4. Stay humble. Stay a learner, and so forth), you can join the dance. Commuters at Grand Central Station suddenly synchronise into a waltz (as happens in Terry Gilliam's *The Fisher King*). Merlin's pupil, Newt, learns, via a series of more-than-human experiences where he plunges into the *Lebenswelt* of different animals, that 'Right is Might', and not its reversal. Harry Potter masters the game of Quidditch. Ged learns The Rule of Names. Lyra Belacqua cracks the alethiometer code. Like the outdoor, active, and kinaesthetic pursuits that helped Meadows formulate her idea ('I had learned about dancing from whitewater kayaking, from gardening, from playing music, from skiing'), all that is required to dance with systems is 'to stay wide-awake, pay close attention, participate flat out, and respond to feedback' (2017 [2004]).

Fantasy, as a manifestation of the world(s) we inhabit and our imaginative participation in their continual interpretation, interpenetration, and transformation, is a complex coalescence of systems—a Qabalistic Tree of Life, an Yggdrasil of multiple worlds—and therefore I think Meadows' approach is wise: 'In a world of complex systems it is not appropriate to charge forward with rigid, undeviating directives' (2017 [2004]). By changing the frame and expanding our time horizons, thought horizons, and 'boundary of caring' (to the more-than-human, and other *Lebenswelts*) like Meadows, we can enjoy the dance.

Crossroads

Finally, it is essential that we acknowledge intersectionality—as writers, readers, consumers, critics, and academics, we do not exist in isolation. Everything we say, feel, and think is inflected by the unique convergence of our gender, ethnicity, socio-economic class, education, politics, religious beliefs, interests, and agendas. We can be self-aware and reflexive about this, and choose to widen our locus of interest, but there is no escaping the essential truth of our 'beingness' on this planet, what is sometimes called our 'positionality'. This is shown here in the idiosyncratic choices of my chosen texts, and style of discourse. Although a healthy cross section is intended, the fallacy of omniscience needs to be acknowledged: no endeavour will be entirely comprehensive or authoritative, as new and neglected work is always emerging. The traditions mutate, and the critical discourse moves on. Rather than seeing this intersectionality as a burden, one can accept the opposite: that it is only as embodied beings firmly upon terra firma that we will rise to the challenges of the Climate Crisis. My argument is that Fantasy enables us to do this, while simultaneously allowing us 'escape, recovery, and consolation', enlarging the creative imaginary of what is possible in doing so. Touchwood, the following chapters will provide enough examples and evidence to convince you of this.

1 Early Fantasy

Green Knights and Faerie Queens

Environmental aspects of Fantasy are entwined with the very roots of the genre, and in this chapter we will consider what John Clute termed 'taproot texts' (1999: 921–922), exceptional works that have significant influence within the genre, including *The Epic of Gilgamesh*, with its destruction of the sacred cedar wood and death of Enkidu, warrior of the wild; *Beowulf* with its marginal outside(r) spaces, the fens, and mere; *Sir Gawain and the Green Knight* and its vivid rendering of the bleak fells and Green Chapel. Yet pre-dating even these proto-Fantasies, you have the Creation Myths of world mythology with their diverse personifications of natural phenomena—flood, thunder, fresh water, hot springs, fertility, and all growing things. The enchantment of nature through narrative (or the acknowledgement of its inherent sacredness) began early in the human project, and (paradoxically) has been entwined with the rise and fall of its civilisations. Ante-Anthropocene Fantasy shows evidence of anthropogenic causality, as in the many Flood Myths resulting from human fallibility. Numerous Fairy Stories and folk tales warn of the consequences of the mistreatment of the natural world, a lack of respect for certain sites associated with otherworldly beings, and the breaking of taboos. Forbidden fruit crops up from the Garden of Eden to the Eildon Hills. Humankind's relationship with nature has been problematic (unsustainable) since the shift from hunter-gatherer to the agrarian at the end of the Mesolithic, and storytelling traditions have reflected that ever since—a kind of collective eco-guilt. The Age of Reason saw a widening of that rift—a rejection of the supernatural in favour of a materialist, mechanistic universe. The separation of Matter and Spirit justified the exploitation of the former in the Old World and the New. Nature became a 'resource'—as did indigenous peoples. Writers of the Enlightenment paid lip service to the quaint notion of enchanted nature—but often merely to entertain. The natural world became framed as 'landscape'—in paintings, plays, and pageants. The world was exhaustively explored and mapped, and the mysterious was relegated to curious marginalia. Addison's 'Fairy Way of Writing' (1712)

DOI: 10.4324/9781003463665-2

was one attempt to redress the balance, as were the Late Victorian Fantasies that emerged against the backdrop of the Industrial Revolution (Richard Jefferies; William Morris; George Macdonald; Fiona Macleod; E.R. Eddison; Lord Dunsany; Andrew Lang), even as Fantasy was relegated to the nursery.

Creation Myths

In a broader sense, *all* myths, legends, folk tales, and Fairy Stories could be considered taproot texts for Fantasy, while acknowledging Sangster's critique that 'Fantasies iterate on things that stand out, but they also rework familiar designs' (2023: 169). Although some key texts indubitably *do* cast a long shadow (e.g. *The Epic of Gilgamesh; Beowulf;* and *Sir Gawain and the Green Knight* are three we shall be considering in this chapter), Sangster argues that the idea of taproot texts should be broadened from individual works to types (2023). Expanding on his analogy of a 'selection of root forms', perhaps taproot texts should be reframed as *mycelial texts*—avoiding the exceptionalism of taproot texts, and acknowledging that the nutrients of influence are rhizomatic: a deterritorialised, complex, entangled biophony, as opposed to the fallacy of a single source. Nevertheless, in the stratigraphy of narrative, myths generally comprise the oldest, deepest layer (with the exception of 'mythopoeic fantasy', as Levy and Mendlesohn term new myth-making texts in 2016: 133–159), and, interestingly, the earliest examples of the Ecological Imaginary.

While applying a taxonomical approach to myths can be useful, it is essential to avoid any claims of cultural equivalence, a Hero's Journey homogeneity. There is no 'monomyth'. A meta-structural analysis identifies common themes in the *creation* of myths, without flatpacking them into reductionist categories that ignore cultural distinctiveness. Only by the close reading of specific myths can salient points be made—and we shall consider a few examples below. But, before we do so, a few 'meta' points are worth making.

To the indigenous culture they emerge from, myths are rarely chronological, as Aboriginal scholar Tyson Yunkaporta argues in his ideation of 'Deep Time Diligence', which responds to the Climate Crisis through a collection application of what he calls TEK (Traditional Ecological Knowledge):

> Deep time diligence is sort of looking at all the systems and the trajectories of these things and doing catastrophic risk analysis, doing all these kinds of things, doing these things *collectively*. So as a group, everybody's out there observing what's happening in nature and what's happening in your economic systems and communities, and we keep coming

> together and everybody's bringing a different data set, and some of these overlap, some of them are contradictory. But in the aggregate, we get a sense, together, with that one big brain—the computational power of a group of people together, you know, big community together doing all this work—*that*, that works. That's deep time diligence, because you start building the stories that you need and the Lore that you need and the knowledge that you need for the system as it's shifting.
>
> (2024)

Myths could be said to exist *before* time, or *outside* the sphere of human time. Myths do not relate, in a literal sense, a series of events that once happened; they are *always happening*. Akin to W.H. Auden's famous, and often misquoted, maxim, that 'poetry does not make anything happen' (he goes onto say it is a 'way of happening'), Creation Myths, which purport to narrativise the roots of the world, are similar. They do not literally describe, even to indigenous cultures (who are never homogenous, but represent a cross section of perspectives from credulity to scepticism), how the world came into being, but are a microcosm of magical thinking. They tell us the world is a storied phenomenon—causal, reciprocal, and symbolic—operating in the poetic language of mythos, as well as logos.

From the first fires to the last fossil-fuelled civilisations, humans have been telling stories (and continue to tell stories) to make sense of the world, to explain how their world, and themselves, have come to be. Myths helped to narrativise catastrophic events, and still do—from storms, earthquakes, and eclipses, to pandemics, terrorist attacks, and Climate Change. We give the unknown agency, demonising anything we cannot predict or empathise with. Whileas now, many turn to the consoling fictions of conspiracy theories to help provide a grand narrative to the multilinear complexity of the world, at one time Creation Myths provided an ontological frame by which to live by.

Rain, thunder, and lightning were turned into surly sky gods, tsunamis and typhoons capricious gods of wave and wind. When volcanoes erupted, or the earth trembled, the gods of the earth were angry. Flood, drought, wildfire, famine, and plague—all were personified as either individual deities or all manifestations of a singular Creator figure punishing its creations. Therefore, Creation Myths are intrinsically 'ecological', as containers for explaining the connections between living systems—because all ultimately relate to the apparent world, its myriad life-forms, topographies, dynamics, and qualities. They do not merely relate how the world came into being, but how to be in the world. Often, they relate how an equilibrium was finally achieved and *how to maintain that equilibrium*: how to maintain good relations with the more-than-human, the gods, spirits, ancestors, and All That Is. Myths are, in the broader sense, fantasies that provide a guide for life. They tell us how to be, or, sometimes, how *not* to be in the world (Manwaring, 2006; Gerdie et al., 2014; Nanson, 2021).

The King and the Wild Man

Inscribed in Akkadian cuneiform onto a dozen fragile clay tablets circa 2000 BCE, and discovered by the Turkish Assyriologist Hormuzd Rassam in the mid-19th century amid the ruins of Nineveh (the ancient city opposite Mosul on the east bank of the river Tigris in northern Iraq), *The Epic of Gilgamesh* is often cited as the earliest, fullest, surviving narrative, and even a proto-novel. It relates the salutary character arc of Gilgamesh, king of the Sumerian city-state of Uruk. In a prologue, he is described as being of both human and divine parentage, an omniscient demigod who, as both fearless warrior and tireless builder, is the father of his embryonic nation. Yet, Gilgamesh's rule is a harsh one, and the sky-god Anu wishes to teach him a lesson. And so Anu creates the king's antithesis, Enkidu, the quintessential wild man, who lives at one in the wilderness with the animals, with whom he can freely converse. If Gilgamesh symbolises civilisation, Enkidu symbolises self-governing nature, or 'self-willed landscape' (Macfarlane, 2007). It seems they are set up as enemies, and yet Enkidu is invited to the city by the king, who learns of this wild man and is intrigued by him, and introduced to its apparent advantages (including the temptations of the flesh—after sleeping with a prostitute Enkidu loses his ability to communicate with the more-than-human). The apparent rivals undergo a trial of strength, which Gilgamesh wins. His dominance established, the king warms to this wild man, and they become friends (companion, or 'servant' in the Sumerian variant). Hearing of the monstrous Humbaba, who fiercely guards a remote, sacred, cedar forest, the two friends set off to destroy the guardian and its groves. Stripping Humbaba of his seven sacred auras, they slaughter him, then spitefully hack down the trees he protected. This sacrilegious act of ecocide ('the first recorded act of human destruction of the natural world,' Flynn and Macfarlane, 2021) accomplished, Gilgamesh returns to Ufuk. Here he stubbornly refuses the hand of Ishtar, the goddess of love, instead choosing to go on another quest with Enkidu, this time to kill a divine bull (sent by the enraged goddess). Enkidu dreams that they have angered the gods by such an act, and he must die. Fulfilling this prophetic dream, the wild man falls fatally ill and dies, leaving the king grief-stricken. The monarch laments his friend, holding an extravagant state funeral. Yet he cannot accept the loss of his dear companion, deciding to undertake a perilous quest to learn from the ancient Utnapishtim, the sole survivor of the Babylonian Flood, the way to conquer death. When he finally reaches the ancient one, Gilgamesh hears the story of the Flood, and is told where to find a sacred flower that will restore his friend to life. Gilgamesh journeys to the farthest shore of a lake filled with deadly water, and there picks the flower. But on the return journey, the exhausted king falls asleep, and a snake devours the hard-won flower of life. Devastated and chastened by this bitter correlative to his arrogance and power, Gilgamesh returns to Ufuk, empty-handed. In an epilogue, the ghost of his friend relates his experiences in the underworld—a place that all the achievements of civilisation could never conquer.

Clearly, *The Epic of Gilgamesh* is a parable about the limits of power and about the necessity of humility in the one's who hold it—a message that resonates over four millennia later. However, in an ecolinguistic reading of the epic, it seems evident that there is an unhealthy balance between the city and the wild—with the ambitions and ethics of the former being antagonistic to the healthy flourishing of the latter. The despoilation of the wild man Enkidu by the sleazy life of the city; the wilful destruction of the sacred cedar wood; the slaughtering of the divine bull; even the plucking of a rare flower—all of these contribute towards Gilgamesh's hard lesson in heartbreak, loss, and the parameters of his own potency. By trying to restore his friend to life (whose demise he catalysed), Gilgamesh is pitching his will against the natural processes of bio-organic life, in which death performs a healthy katabolic function. Civilisation, the ecolinguistic rhetoric of the epic suggests, should not try to work against but *with* nature. The achievements of civilisation can *seem* impressive but, without ecological balance, are ultimately unsustainable. In its many translations and adaptations (e.g. *Lost in the Cedar Wood*, 2021, the Covid-19 'lockdown' album by Johnny Flynn and Robert Macfarlane), *The Epic of Gilgamesh,* speaks to us across the millennia—using the mode of the Fantastic to encourage us to reflect upon the ecological, emotional, and spiritual consequences of our actions.

Although the tablets discovered in the ruins of Nineveh (a poignant, Ozymandian coda to the *Epic*) were not translated until 1861 by the autodidact Assyriologist George Smith, the tropes of the story are universal enough to have manifested frequently in a wide-range of texts: the conflict between modernity and tradition, or the urban and the rural; the assimilation or rehabilitation of the 'wild twin'; the slaying of the monster; the quest for the elixir of life; bereavement narratives and the acceptance of mortality; and an arrogant, powerful man is taught a painful lesson in humility—all tropes that resonate down through the ages. The early provenance of *The Epic of Gilgamesh* suggests it could be narrativising the painful transition from a hunter-gatherer to an agrarian culture. This took place at the threshold between the Mesolithic and Neolithic (circa 2700 BCE, approximately 700 years before the composition of the epic, but a major paradigmatic shift conceivably retained in the oral tradition until finally written down). Thus, it could be seen as a proto-text for one of many describing our changing relationship with the land. What interests us here in particular are the fantastic elements—the supernatural consequences for an unhealthy imbalance, or downright abuse, of the biome we share with the more-than-human. This trope manifests in our next 'mycelial text'.

The Hero and the Monster

The anonymous Anglo-Saxon poem, *Beowulf*, comprises approximately 3182 lines and was composed sometime between 700 CE and 1000 CE. The plot is familiar to many, not least to Tolkien, who, as a philologist and Fantasy author, was heavily influenced by it (his translation was considered to be the finest for many years). His vision of the Rohan—the proud king in his hall, the malcontent adviser, the high and mighty horse-lords and brave shield-maidens, the clan besieged by a monstrous threat, the power of poetry, and importance of honour and glory in the face of death—all remediate the poem within a High Fantasy setting. Neil Gaiman's adaptation for the big screen (dir. Zemeckis, 2007) brought its attention to a wider audience, although that version re-imagined the story for a 21st-century audience. John Gardner's *Grendel* (1971) retells the story from the perspective of the titular monster, who Gardner casts as an archetypal malcontent outsider (satirically modelling him upon the existential philosopher, Jean-Paul Sartre). In an innovative reconfiguration, Gardner dramatises conversations between Grendel and the nameless dragon who will one day avenge the monster-slaying Beowulf. In *The Mere-Wife* (2018), Maria Dahvana Headley reimagines *Beowulf* in a contemporary American setting, focusing on the experience of a human version of Grendel's mother, a traumatised marine, who escapes from her captivity and ends up living near 'Herot Hall', a new development named after Roger Herot, a plastic surgeon and heir to the Herot estate. She ends up marrying him, but before giving birth to her son, Gren—possibly the product of rape. Gren proves to be 'monstrous', but no less than his human peers. Nonetheless, he manages to befriend one, Dylan. Beowulf appears as police officer Ben Woolf, who goes searching for Dylan when the boy goes missing. Things play out tragically, in the gritty realism of a dark crime story—perhaps reflecting more the zeitgeist of Noughties' America, a nation traumatised by 9/11, and caught up in the Gulf War conflict and divisive rhetoric that followed, than anything in the source text. Yet, Headley's novel, and the many other adaptations produced over the last century, show the remarkable plasticity and universality of the story.

In the original text, Beowulf is a warrior of renown from Geatland, who is summoned by the neighbouring king, Hrothgar, whose hall, Heorot, is assailed every nightly by a blood-thirsty creature of the fens, Grendel. This assault has continued for twelve years. Using himself and his warriors as bait, Beowulf sleeps in the hall, and manages to fight off an attack, ripping off Grendel's arm in the process. Yet, it is clear the threat is not yet neutralised—for Grendel has only retreated to the lake in which dwells his monstrous mother. Beowulf valiantly pursues the enemy into its lair, and despatches both, returning with the head of the mother. Feted as a hero by a grateful Hrothgar, Beowulf grows rich and powerful, and eventually takes on the throne he has saved. He reigns successfully for fifty years, until a thief steals a goblet from a dragon's hoard, which stirs it into action, laying waste to the kingdom. Beowulf decides he must take on this 'one last job', despite being advanced in years. He journeys to the lair, contemplating upon his life, on glory, and on the emptiness of wealth. In a fierce battle he destroys the dragon, but at the cost of his own life. He is laid to rest, and lauded as a great hero, but one whose earthly accomplishments are put into perspective by a Christian gloss.

Let us now consider *Beowulf* environmental aspects as they manifest as themes and motifs. The most common imagery is that of sunrise and night. 'Dawn' and 'midnight' recur several times, for example, 'Then out of the night/came the shadow-stalker, stealthy and swift' (lines 702–3); and 'when morning light,/scarfed in sun-dazzle, shines forth from the south/and brings another daybreak to the world' (lines 603–6).

These contrapuntal strands interweave the whole text like knotwork. This diurnal imagery suggests a Manichaean struggle. The light is associated with good, the dark with evil. Beowulf in all his shining war-gear and warrior bling is the solar hero. His destiny is to save shining Heorot from the forces of night—Grendel, Grendel's mother, the creatures of the mere, and the dragon—the forces of chaos and destruction. The fen-dwelling Grendel is a creature of liminal spaces—he brings the cthonic, nocturnal energies into the bright hall of Hrothgar:

> '[Beowulf] realized/that the demon was going to descend on the hall,/ that he had plotted all day, from dawn-light/until darkness gathered again over the world/and stealthy night-shapes came stealing forth/ under the cloud-murk.'
>
> (Heaney trans., 2000: 22)

The main dialectic of the poem is between civilisation and wilderness, continuing the themes of *The Epic of Gilgamesh*. Heorot is the new Uruk; Hrothgar the new Gilgamesh; but now both Beowulf and Grendel contain elements of the wild man Enkidu—both come from 'outside', and embody primal traits and great strength. Grendel and Grendel's mother are, collectively, Humbaba. The dragon is the divine bull. In *Beowulf,* civilisation is symbolised by the shining hall of Heorot, the pinnacle of the warrior culture of the period ('the wonder of the world forever' 2000: 5/line 71). It metonymically represents societal order, hierarchy, and the protection of a strong leader. It is a contained but collective space:

> 'They marched in step,/hurrying on till the timbered hall/rose before them, radiant with gold./Nobody on earth knew of another/building like it. Majesty lodged there,/its light shone over many lands.'
>
> (2000: 12)

In contrast, the wilderness is open country, uncontrolled, and, at that point, unenclosed and illimited. The sea, fens, mere, and dragon's lair symbolise wilderness, and the more-than-human. They are organic and natural, even if filled with dread and danger to humans. Even the mere, seething with monsters, is full of life. Yet, in contrast to the safety in numbers of the human tribal system, Grendel, Grendel's mother, and the Dragon all seem to lead lonely, isolated lives, disconnected from community:

> 'Grendel was the name of this grim demon/haunting the marches, marauding round the heath/and the desolate fens; he had dwelt for a time/in misery among the banished monsters,/Cain's clan, whom the Creator had outlawed/and condemned as outcasts.'
>
> (lines 102–107)

An ecolinguistic reading of the poem would suggest that exploitation of natural resources (by Hrothgar and his realm), such as the felling of timber for Heorot, and the plundering of precious ore to make the much-prized gold, leads to consequences. Tellingly, this wealth is either destroyed (Heorot is nearly torn apart) or returned to the ground (the dragon's hoard is buried with Beowulf). This may suggest a Christian subtext, about the vacuity of materialism. Riches are ultimately hollow—you cannot take them with you. The anonymous poet seems to be saying: attend to your soul—fix your gaze on the next life. For a warrior culture this is primarily through the notion of honour, but it is an anti-materialist discourse. The way one governs oneself in battle, in the mead hall, is more important than any trappings of wealth. One's name and one's word are priceless. Bravery is worth more than gold. Steadfast Wiglaf is worth more than twenty men. One's deeds live on beyond you. Only your reputation remains.

Beyond these themes and motifs there is a tangible sense of *place* throughout the poem, of regionality, seasonality, and placiality (to use Casey's term). This sense of situatedness is there from the outset. In the second stanza we hear, 'In the end each clan on the outlying coasts / beyond the whale-road had to yield to him / and begin to pay tribute' (Heaney, trans., 2000: 3). The littoral communities are connected by geographical kennings ('whale-road'; 'gannet's bath', etc.) that show the familiarity of this elemental infrastructure, the binding solution of seafaring people.

Macrocosmically, the world of Beowulf is one created by a divine intelligence who has ordered the universe in a meaningful way. In a scene echoing the Christian Genesis, a scop in the mead hall relates this Creation:

> 'the clear song of a skilled poet/telling with mastery of man's beginnings,/how the Almighty had made the earth/a gleaming plain girdled with waters;/in His splendour He set the sun and the moon/to be earth's lamplight, lanterns for men,/and filled the broad lap of the world/with branches and leaves; and quickened life/in every other thing that moved.'
> (2000: 5/lines 90–98)

The scop's performance is an instantiation of this fundamental act of creation, an echo of the primal breath upon the waters, and the Word that made the world. This is reified by the original context of *Beowulf*—performed as an oral text from memory within the mead hall. This has been recreated by the actor Julian Glover, performing the text in the original Anglo-Saxon in a reconstruction of a Viking longhouse (2008); by the East Anglian storyteller Hugh Lupton at the Beyond the Borders International Storytelling Festival (2005); and myself in a roundhouse at the Peat Moors Visitor Centre on the Somerset Levels (2000); and virtually by the Icelandic author and tour guide Svanur Gisli Thorkelsson, for Audible (2020). What Finnish philologist and compiler of the *Kalevala* (his 'national epic', with analogous oral and environmental qualities to *Beowulf*) Elias Lönnrot calls the 'performative' text (Honko, 2002) is a fluid, customisable state until it is finally written down. Until then, each performance has its own distinctive environmental qualities (Manwaring, 2006; Gerdie et al., 2014; Nanson, 2021), which extend beyond the text on the page: the acoustic; ambience; temperature; aesthetic of the venue; presence of the audience; and, if outside, the weather and local flora and fauna—all of which create a subtle phenomenological feedback loop enriching that unique performance. This is the wider ecological quality of any text that is performed, as many of these examples of Early Fantasy were. This reaches its apotheosis (in the West, at least) in the courts of Medieval Europe, where oral tradition fused with the burgeoning literary forms of the Gest, Ballad, Lay, and Romance.

The Court and the Wilderness

Sir Gawain and the Green Knight is an alliterative poem, or lay—a narrative poem designed to be sung, and 'usually dealing with matter of history or romantic adventures' (Drabble, 1985: 556)—composed in the Late 14th century by an anonymous poet writing in a north-west Midlands variant of Middle English. Comprising 2530 lines, it is divided into four narrative 'fits'. It is a hugely influential poem, being much translated, adapted, and reworked into fine art, fiction, theatre, opera, graphic novels, roleplaying games, tarot cards, and music. This is all the more remarkable, considering it survives in a single, fragile, manuscript, Cotton Nero A.x (which also includes the Medieval poems, *Pearl, Cleanness,* and *Patience*), which was nearly destroyed by a fire in the library of antiquary Sir Robert Bruce Cotton in 1731.

In his 1953 W.P. Ker Memorial Lecture on *Sir Gawain and the Green Knight,* Tolkien defends it as 'a poem that deserves close and detailed attention' (2021: 109). He argues for its literary merits and moral instruction: 'It is made of tales often told before and elsewhere, and of elements that derive from remote times, beyond the vision or awareness of the poet' (2021: 109). This acknowledgement of intertextuality refutes the notion of singular 'taproot texts', and supports my notion of 'mycelial texts': that Fantasy flourishes within an ecosystem—the more diverse the better. Nevertheless, Tolkien uses the term, '*rooted* works' (2021: 109), and 'a real deep-rooted tale' (2021: 110), suggesting how they flourish precisely because of the narratological tree from which they are nurtured: 'Behind our poem stalk the figures of elder myth, and through lines are heard the echoes of ancient cults, beliefs and symbols remote from the consciousness of an educated moralist (but also a poet) of the late fourteenth century' (2021). The poem receives nutrients from the earlier sources: 'it receives part of its life, its vividness, its tension, from them' (2021). Its *viriditas* (to use Hildegard of Bingen's term for the 'greening of the soul') draws upon all of this. Yet, as with LOTR, the spiritualised ambience is certainly not purely a Christian one. There is the rarified atmosphere of *Faerie* infusing this specifically northern landscape, and transmogrifying the actuality of nature into something else—as symbolised by the 'greener than it seemed' (stanza 11) hue of The Green Knight and his armour, steed, and axe, as Tolkien emphasised: 'If we are introduced to a green man, with green hair and face, on a green horse, at the court of King Arthur, we expect "magic"' (2021: 113).

The plot of the poem goes as follows. During the Yuletide revelries at Camelot, King Arthur insists that the Court hears of a marvel before the feast can begin: 'of some strange story or stirring adventure,/or some moving marvel that he might believe in' (stanza 5). Just then the doors of the hall burst open and an uncanny knight, arrayed from head to toe in green, upon a green steed, enters: 'as a fay-man fell he passed, and green all over glowed' (stanza 7). He bears a holly branch and an axe. The Green Knight challenges any to strike a blow to his naked neck—if they agree to receive one in return. Arthur initially steps up, but his nephew, Gawain, persuades his uncle that he should take his place, being the least accomplished knight and lowliest there, except by favour of his royal connection. Arthur agrees, and Gawain doughtily smotes off the head of the Green Knight. However, the head is picked up by the decapitated knight, and speaks—providing details of the reciprocal arrangement—in twelve months' time at the Green Chapel, somewhere in the wild north. The Green Knight departs, head in hand. Gawain is congratulated for his bravery, and the revels continue—but for a full year Arthur's nephew is aware of the fate hanging over his head. As midwinter approaches, Gawain sets off, seeking the Green Chapel. He has no luck, and the weather is getting worse, when he stumbles upon a fine-looking castle. There he receives a warm welcome from the merry Lord and his comely Lady. Gawain is invited to stay over Yuletide until New Year, for the Green Chapel is only a couple of miles away. The Lord insists on a game whereby they exchange the 'winnings' of their day respectively. Gawain agrees. The Lord goes off to hunt, returning on the first day with a deer, then a boar, and then a fox. Each day the Lady comes to Gawain's chamber, makes flirtatious love-talk, and insists upon a kiss—these the knight offers to the returning Lord, but does not reveal their provenance, as their covenant did not agree to that. However, on the final day the Lady also offers Gawain her green silk girdle—saying it has magical properties that will protect him from the fatal blow that awaits him. Gawain does not disclose this to the Lord. He bids his hosts a farewell, and rides to the Green Chapel. His local guide tries to dissuade him, but Gawain pushes on alone. The Green Knight awaits, axe ready. He feints twice, and then nicks Gawain's neck on the third—this is to mirror the kisses shared honestly, and the concealment of the green garter. The Green Knight considers their agreement honoured. Revealing his name to be Lord Bertilak, he explains he has been enchanted by Morgan Le Fay (Gawain's aunt) into the Beheading Game as a way of humiliating the Court, in particular Queen Guinevere. Chastened, Gawain departs, and returns to Camelot, where he relates his whole adventure to the Court. Arthur and his knights agree to all wear a green sash in solidarity with Gawain, and in acknowledgement of their *own* fallibility.

Now, let us consider the main symbolic binary here between the Court and the wilderness. The former represents 'civilisation' in microcosm: wealth, opulence, bounty, but at the expense of others' hard work (a feudal system, tithes, serfdom, etc):

> 'This king lay at Camelot at Christmas-tide
> with many a lovely lord, lieges most noble,
> indeed of the Table Round all those tried brethren,
> amid merriment unmatched and mirth without care.'
>
> (stanza 3)

Although this could be seen as a symbol of unity, of democracy, and the nobler aspects of human nature (as indeed it was for the Kennedy

Administration, who adopted the notion of 'Camelot'), a Marxist view would interpret the scene very differently. It is one of extreme privilege: the 'mirth without care' is due to the hidden labour of others—the servants, who are literally reduced in size (and thus importance) in Medieval paintings of courtly banquets.

In contrast to this stands the Green Knight and the Green Chapel, who between them symbolise the more-than-human, the supernatural. Here, the 'wild man' (a Medieval Enkidu) is more knightly, and can conduct himself in a chivalric way (later we discover he is Lord Bertilak under enchantment). He has not only learnt the rules of the Court, he can create his own. From Humbaba to Grendel to the Green Knight—we can see a significant shift. The wild, it seems, is being civilised, or perhaps placed on a more equal footing. Certainly the Green Knight is more than a match for any in the Court, and does not kowtow to King Arthur—he is very much in touch with his own sovereignty: he is in effect *Rex Nemorensis*, King of the Wood (as the hunting scenes later demonstrate). The poet emphasises the Green Knight's primal power:

> 'when there passed through the portals a perilous horseman,
> the mightiest on middle-earth in measure of height,
> from his gorge to his girdle so great and so square,
> and his loins and his limbs so long and so huge,
> that half a troll upon earth I trow that he was,
> but the largest man alive at least I declare him;
> and yet the seemliest for his size that could sit upon a horse'
> (stanza 7)

The telling detail of his shoelessness 'for shoeless he rode,' (stanza 8) suggests a connection to the earth, and to the Earth. The holly bush he wields is a symbol of peace; the axe, in contrast, is a man-made weapon of violence (to nature; to humans):

> 'And yet he had not a helm, nor a hauberk either,
> not a pisane, not a plate that was proper to arms;
> not a shield, not a shaft, for shock or for blow,
> but in his one hand he held a holly-bundle,
> that is greatest in greenery when groves are leafless,
> and an axe in the other, monstrous and ugly.'
> (stanza 10)

The Green Knight reassures the Court that he does not come to fight them: 'You may believe by this branch that I am bearing here/that I pass as one in peace, no peril seeking' (stanza 12). Yet the holly bush could also symbolise a kind of truce between the human and more-than-human worlds.

And yet there is something distinctly uncanny about the Green Knight:

> 'Then they looked for a long while, on that lord gazing;
> for every man marvelled what it could mean indeed
> that a horseman and horse such a hue should come by
> as to grow green as the grass, and greener than it seemed,
> than green enamel on gold glowing far brighter.'
>
> (stanza 11)

This is the fantastic in the service of the environment—'some moving marvel that [we] might believe in' (stanza 5)—a prime example of the Ecological Imaginary. The Court are not in doubt as to the supernatural provenance of the visitor before them—'wherefore a phantom and fay-magic folk there thought it' (stanza 11)—nor to its reality—'a marvel beyond denial' (stanza 20)—and immediate threat to the safety and complex etiquette atmosphere of the Court. Such a primal incursion (an early example of Intrusion Fantasy) is a gauntlet to the highly artificial societal fabric, a rude reminder of all that such artifice risks losing if it disconnects too much from the raw power of nature. This was Gilgamesh's lesson, and, three and half millennia later, it is King Arthur's.

The Green Knight's challenge seems to be designed to elicit the inner wildness out of the knights, stating that the candidate for the Beheading Game would be suitable: 'if so bold be his blood or his brain be so wild' (stanza 13). He is emphasising a different kind of consciousness and mien—one that is more embodied and connected to true feelings, and to the natural world. Once the blow is delivered—a moment of visceral gore, that brings the horror of the hunt (the butchery as described later) into the civilised Court, the Green Knight picks up his head and reminds Gawain of their pact:

> 'To the Green Chapel go thou, and get thee, I charge thee,
> such a dint as thou hast dealt—indeed thou hast earned
> a nimble-knock in return on New Year's morning!'
>
> (stanza 20)

The karmic wheels have locked into place, and Gawain's destiny is synchronised with the wheel of the year. The poet emphasises this with vivid descriptions of the seasons:

> '. . . then the weather of the world makes war on winter,
> cold creeps into the earth, clouds are uplifted, shining
> rain is shed in showers that all warm fall on the fair turf,

> flowers there open, of grounds and of groves green is the
> raiment, birds are busy a-building and bravely are singing
> for sweetness of the soft summer that will soon be on
> the way.'
>
> (stanza 22)

The wheel of the year seems to mirror the wheel of fortune—the inexorability of Fate:

> 'all things ripen and rot that rose up at first,
> and so the year runs away in yesterdays many,
> and here winter wends again, as by the way of the world
> it ought.'
>
> (stanza 23)

This is ecological karma in action—anthropocentric accountability. The only acceptable payment is a full-bloodied commitment, which is literalised by Gawain's own blood spilling onto and metaphorically fructifying the soil, a symbolic repayment to the Earth. This occurs in the finale of the Beheading Game at the Green Chapel, when, faced with his fate ('beheaded by an elvish man for an arrogant vaunt' (stanza 29)), the Green Knight's axe nicks Gawain's neck ('so that the shining blood o'er his shoulders was shed on the earth' (stanza 93)).

Yet, before this, we have Gawain's long and arduous journey in search of the site of ecological justice, the natural court of the Green Chapel. The poet describes in great detail Gawain's accoutrements, mirroring the description of the Green Knight but contrasting it with the artifice of Camelot, where rank, status, and prestige is shown by one's attire, war-gear, and horse: 'Now he rides thus arrayed through the realm of Logres' (stanza 30). Yet once Gawain leaves Court, none of these trappings matter beyond their practical value in the wild: 'He had no friend but his horse in the forests and hills' (stanza 30). He is forced to take a route away from all that he knows into a terra incognita beyond the pale of the Court:

> 'The knight took pathways strange
> by many a lonesome lea,
> and oft his view did change
> that change ere he could see.'
>
> (stanza 30)

In effect, Gawain rides to the edge of the map, where he not only finds monsters but must contend with them, thus firmly placing him in the realm of the fantastic and foreshadowing many popular tropes of Tolkienian Fantasy:

> 'At whiles with worms he wars, and with wolves also,
> at whiles with wood-trolls that wandered in the crags,
> and with bulls and with bears and boars, too, at times;
> and with ogres that hounded him from the heights of the fells.'
>
> (stanza 31)

Here, they are summarised in four lines, but later writers of Sword and Sorcery, and High Fantasy, spin whole novels, indeed whole series, from such a litany of the monstrous.

However, it is not only the *fauna* that is antagonist, but the landscape and elements themselves:

> 'the winter was worse,
> when the cold clear water from the clouds spilling
> froze ere it had fallen upon the faded earth.
> Wellnigh slain by the sleet he slept ironclad
> more nights than enow in the naked rocks,
> where clattering from the crest the cold brook tumbled,
> and hung high o'er his head in hard icicles.
> Thus in peril and pain and in passes grievous
> till Christmas-eve that country he crossed all alone
> in need.'
>
> (stanza 31)

In such scenes, Gawain is far from the glamour of Court. In the wild he must draw upon his own resources, and is faced with the tangible consequences of his actions. Bound by them, he sleeps 'ironclad', locked into his decision to play the Beheading Game. Searching for an unknown location, it seems Gawain must embrace Keatsian 'negative capability'. Lost in mists on lonely moors, it is as though the Cloud of Unknowing has descended. He experiences the katabasis of the dark night of the soul before 'seeing the light'. Mercifully, on Christmas Eve, Gawain's prayers are answered, and he beholds Bertilak's castle, a most welcome refuge in the wilderness:

> 'That brave bridge was up-hauled,
> the gates were bolted fast;
> the castle was strongly walled,
> it feared no wind or blast.'
>
> (stanza 33)

Here, the beleaguered Gawain receives a warm welcome: 'All hasped in his harness to hall they brought him, / where a fair blaze in the fireplace fiercely was burning' (stanza 35). The Lord of the castle greets Gawain by bending his knee to the cold earth, and is described as having 'a face fell as fire' (stanza 36), and yet he is merry is demeanour and is a doughty, manly fellow, associated with redness—'bright was his beard, and all beaver hued' (stanza 36)—in vivid juxtaposition to his alter-ego, the Green Knight.

Once Gawain is allowed to recover from the ordeal of his winter journey, the Lord suggests his intriguing agreement, which, in the famous 3rd fit, tellingly juxtaposes the daily hunting with the 'hunting' of the bedroom, in which the Lady of the Castle tests the young knight's courtesy and resolve:

> 'Thus in his game the lord goes under greenwood eaves,
> and Gawain the bold lies in goodly bed.'
>
> (stanza 48)

The convincingly detailed descriptions of hunting suggest that the poet had some familiarity with the hunt. The sequence of deer, boar, and fox could suggest something heraldic to the Medieval audience or reader. However, the quarry pursued by the indefatigable Bertilak and his hunt seem to be more than merely allegorical. In stanzas 62–64 in particular, the description of the boar-hunt appears to emphasise the traumatic experience from the animal's point of view. This seems to be an attempt at empathy with the more-than-human. The boar is given a character, agency, and, thus, sentience. It struggles to stay alive, and is fierce until the bitter end. It is echoed by John Clare's poem, 'The Badger'. The same non-anthropocentric sympathy occurs with the fox-hunt (stanzas 68–69), in which the pursued fox uses his 'wiles' to try and escape the hunters and their hounds:

> 'He swerves then swift again,
> and dauntless darts astray;
> in grief and in great pain
> to the wood he turns away.'

Each hunt ends with the spoils of nature being divided up, and yet there appears to be a law of diminishing returns, for the last is dismissively described by Bertilak as 'this foul fox-fell' (stanza 77). The contrast with the *increase* in Gawain's daily love-game gains is poignant (Gawain is the quarry, though he does not realise it), but ecologically speaking, this could signal biodiversity loss to a modern reader—the consequences of man's rapaciousness.

Antithetically, it could be argued that the hunting scenes actually dramatise a deep knowledge of the land, and its flora and fauna—not an eco-illiterate ignorance, or mindless exploitation of it. The hunter is especially sensitive to the habitat and breeding cycles of his prey. It requires close observation and biophilic attunement: 'The man is aware of the wild thing, and watchful awaits him' (stanza 76).

Whatever one thinks of that, there is a definite archetypal quality to the patterning of the 3rd fit. The exchange of the spoils of the hunt for kisses could suggest a Thanatos–Eros dynamic—love conquers death, or at least holds it at bay. The two are held in a mutualistic tension.

This queasy line between desire and fear runs like a thread throughout the poem—from the push/pull factors that lead to Gawain volunteering himself for the Beheading Game, to the perilous love-games of Bertilak's castle. Maintaining one's honour is weighed against the desire to stay alive: morality and mortality are the counterweights of this text.

Transcending these are the primal forces of creation and destruction, which bind the whole of nature. It seems our protagonist is forced to walk a knife's edge between the two. Reluctantly leaving Bertilak's castle on New Year's Day (when the pathetic fallacy of the dreich weather makes getting out of bed on his day of reckoning a drear prospect), Gawain's guide leads him to the 'grievous place/where he is due to endure the dolorous blow' (stanza 83). The poet describes their hard journey to the Green Chapel: a classic scene of Early Fantasy, emulated in countless Fantasy stories, where the landscape seems animate, malignant, and complicit:

> 'They go by banks and by braes where branches are bare,
> they climb along cliffs where clingeth the cold;
> the heavens are lifted high, but under them evilly
> mist hangs moist on the moor, melts on the mountains;
> every hill has a hat, a mist-mantle huge.
> Brooks break and boil on braes all about,
> bright bubbling on their banks where they bustle downwards.
> Very wild through the wood is the way they must take,
> until soon comes the season when the sun rises
> that day.'
>
> (stanza 83)

The descriptions of the landscape through which Gawain passes feel so real—suggesting experience of actual places, journeys, etc. This is reflected in the ecolectical vocabulary of the Gawain poet, which suggests a northern provenance, as noted by many Gawain scholars, and Simon Armitage, in his 2008 translation, who said, 'coaxing Gawain and his poem back into the Pennines was always part of the plan' (interviewed by Lezard, *Guardian*, 2008).

At the threshold of the Green Chapel, the nameless retainer of Lord Bertilak does his best to discourage Gawain from entering, as though to test his resolve: 'The place that you pass to, men perilous hold it, / the worst wight in the world in that waste dwelleth' (stanza 84). To emphasise the point, he offers a cowardly Parthian Shot, 'For all the gold in the world I would not go with thee . . .' (stanza 86). Nevertheless, Gawain grimly continues to the destination of his destiny, a bleak gorge at the head of which is situated what sounds like a Neolithic burial chamber (stanza 87), a prehistoric monument which was frequently attributed to The Devil (e.g. The Devil's Den in Wiltshire). Here, the chilling sound of the Green Knight sharpening his axe echoes against the rock (stanza 88), like the Grim Reaper whetting his scythe. This is the moment of reckoning. At first Gawain flinches, and the Green Knight mocks him, but steeling himself, the young knight stoically stands, awaiting his fate. It is telling that, in this moment, Gawain almost becomes as natural as the Green Knight:

> 'Gawain warily awaited it, and winced with no limb,
> but stood as still as a stone or the stump of a tree
> that with a hundred ravelled roots in rocks is embedded.'
> (stanza 92)

And in such a way, Gawain passes this deadly test, receiving only an admonitory nick in the neck for his concealment of the Lady's girdle. Immediately, the Green Knight changes demeanour, becoming merry and revealing himself to be Lord Bertilak and the orchestrator of all the challenges Gawain has faced since arriving at his castle:

> 'Now I am aware of thy kisses, and thy courteous ways,
> and of thy wooing of my wife: I worked that myself!
> I sent her to test thee, and thou seem'st to me truly
> the fair knight most fautless that e'er foot set on earth!'
> (stanza 95)

Nothing is hidden from Bertilak. Gawain is mortified, but the Lord reassures him his fallibility is merely mortal, and the weakness of his sex: 'But no marvel it is if mad be a fool,/and by the wiles of woman to woe be brought' (stanza 97).

So, it is fitting that Bertilak allows Gawain to retain the 'sovereign device' (stanza 74) girdle of the Lady of the Castle, as a metonymic reminder of her test of his 'greenness', the Green Knight, and the Green Chapel:

> 'And I give thee, sir, the girdle with gold at its hems,
> for it is green like my gown. So, Sir Gawain, you may

think of this our contest when in the throng thou walkest
among princes of high praise; 'twill be a plain reminder
of the chance of the Green Chapel between chivalrous knights.'
(stanza 96)

The Court and the Wild are united via the presence of the green girdle, which encompasses both. Thus, these two 'worlds' are woven together, and the dialectical tensions of the poem are reconciled. Wearing this 'token of [his] trespass' (stanza 98), Gawain returns to Camelot, humbled, and yet, strangely empowered, for forthwith, the young knight embodies the wild—a proto-Green Knight: 'Wild ways in the world Wawain now rideth' (stanza 100). Having been the opponent of the wild, he is now its ambassador.

These ecological themes are teased out even more in the iconoclastic David Lowery film adaptation—*The Green Knight* (2021). In Lowery's dreamlike, visionary treatment, the titular antagonist is portrayed as the Green Man: not just green-skinned, but bark-skinned, like a wodwose, or the foliating head of many an English church. Gawain experiences an eco-epiphany in the final encounter. Instead of a violent reckoning, Gawain arrives to find the Green Knight sitting in the niche of a ruined chapel. Gawain sits at his feet in Buddha-like contemplation. Finally, at dawn, the Green Knight stirs, and the Beheading Game resumes. An alternative ending is played out in which Gawain flees his fate, takes a wife, and becomes a cruel-hearted king. Earlier, another quantum divergence is depicted: Gawain is depicted as being bound and left to die in an ambush—his shield bearing the ikon of the Virgin Mary broken, his horse stolen—his corpse slowly becoming a bleached skeleton as the seasons turn. Then the camera pans back, the seasons reverse, and Gawain revives and resumes his journey. A talking fox guides Gawain, and apparently peaceful giants stride the land with no explanation. All of this is situated within a very tangible northern landscape.

The overlapping and intermingling of fantastical and naturalistic elements are discussed in the final section of this chapter.

Enchanted Otherworlds

In Early Fantasy, the Otherworld (Faerie; Annwn; Tir nan Og; Elfhame; Lyonesse; Avalon) is often close by—accessible by accident, invite, boldness, or fate—and sometimes expansive, or co-mingling. Its sense of proximity, of it existing parallel to our world in a perilous crosshatch chanced upon at hawthorn trees, fairy rings, springs, pools, bridges, caves, crossings, dusk, midnight, dawn, May Eve, Midsummer, or Samhain, suggests a kind of Palimpsest Fantasy—in which multidimensional ontological layers, or lifeworlds, overlap. Relations between these worlds can be playful, formal, beneficial, parasitic, frozen, or easily damaged by thoughtless actions, such as the felling of a fairy tree, or the breaking of some taboo or geasa. In terms of the Ecological Imaginary, this creates a sense of immanence in the natural environment—as though at any moment a deeper reality or marvel is about to break through. Proximity to Faerie is foreshadowed by a thickening of atmosphere, ethereal music, an effulgence, a heightened awareness, and an intensification of the metaphoric freight of everything perceived. This ensouled landscape can make us reconsider the actuality of the world around us. The next time we go for a walk in a wood, we may imagine ourselves stepping into Faerie, or encountering a being from another (or co-existent, but hidden) reality. Palimpsest Fantasy also emphasises the consequences of our actions—as in the accelerated karma and narrative compression of a story, rash, thoughtless behaviour leads to dramatic, immediate impact, the aftermath of which sometimes taking years to work through or unravel. In such an intensely causal environment, the right etiquette is essential—the importance of manners and correct forms of address, the significance of names, titles, and protocol, and actions undertaken in the right order, at the right time, with a scrupulous attention to detail—all of these things could be seen as highly mannered, and a mock-show of the upper classes. However, they can also be read as a form of animism, in which everything has sentience and agency, and needs to be duly honoured with the correct rites and offerings—as practised in numerous indigenous cultures around the world. The ecolectical specificity of stories from oral traditions (collected, transcribed, edited, and rewritten by Victorian collectors) suggests a formalised indigeneity, which although connected to the nation-building projects of the 19th century, can be reclaimed today in a more fluid, transnational way, as a form of ecoliteracy: an attempt to listen to, speak to, and respect the other-than-human.

Examples of immanence, consequence, and etiquette can be seen in many tales from Celtic myths and legends. One of the most significant collections of material is in *Y Mabinogi* (The Mabinogion), twelve tales from the Welsh bardic tradition compiled by Lady Charlotte Guest in 1877 from *The White Book of Rhydderch* (c.1300–1325) and *The Red Book of Hergest* (1375–1425), along with the tale of Taliesin, taken from a 16th-century manuscript (*Hanes Taliesin*). In the latter, a witch called Ceridwen gives birth to an ill-favoured son, Afagddu—to compensate, she decides to concoct a potion of inspiration for him. To prepare it she gets an old man to maintain a cooking fire, and a young boy called Gwion Bach to stir it for a year and a day. The boy drops the spoon in the potion, and it scalds his hand. Without thinking, he puts his burnt fingers in his mouth and receives the wisdom, distilled into three drops, meant for Afagddu. When Ceridwen discovers this, she is furious and pursues the boy. However, he now has the shape-changing power of fith-fath and turns into a hare; Ceridwen turns into a greyhound; he turns into a salmon; she turns into an otter; he turns into a wren; she turns into a hawk; finally, he turns into a grain of wheat, hiding on the threshing floor of a farm, but she turns into a hen and gobbles him up. Nine months later, she gives birth to a boy glowing with awen (Welsh, noun: inspiration). She cannot keep it and casts it beyond the ninth wave. It is found by Elphin, caught in a salmon weir. When he spots the child, he cries out, 'Behold, Taliesin!' (the radiant brow). He and his father raise him as a foster child. This wunderkind has the power of speech and the gift of prophecy, maturing into a youth of uncanny abilities. He is taken to the court of a powerful ruler, where he is tested: proving himself to be the greatest of bards, he goes on to fulfil his destiny as 'penbeirdd', chief-bard of Wales, and of King Arthur's court. Although a semi-legendary figure, a substantial corpus of poems and prophecies is attributed to him. The fith-fath in this story is mirrored by tales of anamorphic shamanism across the world—more closely to Wales, in *The Song of Amergin*, across the Irish Sea—which suggests a kind of more-than-human supra-consciousness:

Amergin, Bard of the Milesians, lays claim to the Land of Ireland

I am a stag: of seven tines,
I am a flood: across a plain,
I am a wind: on a deep lake,
I am a tear: the Sun lets fall,
I am a hawk: above the cliff,
I am a thorn: beneath the nail,
I am a wonder: among flowers,
I am a wizard: who but I
Sets the cool head aflame with smoke?

I am a spear: that roars for blood,
I am a salmon: in a pool,
I am a lure: from paradise,
I am a hill: where poets walk,
I am a boar: ruthless and red,
I am a breaker: threatening doom,
I am a tide: that drags to death,
I am an infant: who but I
Peeps from the unhewn dolmen, arch?

I am the womb: of every holt,
I am the blaze: on every hill,
I am the queen: of every hive,
I am the shield: for every head,
I am the tomb: of every hope.
(trans. Robert Graves, 1961)

In the Celtic Tradition, animals are frequently harbingers and signifiers of the Otherworld. In the first of *The Four Branches of the Mabinogi*, 'Pwyll Prince of Dyved', the titular prince is out hunting, pursuing a stag, when it is run to ground by a pack of fairy hounds:

> And he beheld a glade in the wood forming a level plain, and as his dogs came to the edge of the glade, he saw a stag before the other dogs. And lo, as it reached the middle of the glade, the dogs that followed the stag overtook it and brought it down. Then looked he at the colour of the dogs, staying not to look at the stag, and of all the hounds that seen in the world, he had never seen any that were like unto these. For their hair was of a brilliant shining white, and their ears were red: and as the whiteness of their bodies shone, so did the redness of their ears glisten.'
> (2000: 3)

Whiteness is a symbol of the Otherworld (Annwn; Annwvyn) in *The Mabinogion* (Guest, 2000). The hounds belong to Arawn, Lord of Annwvyn, the Celtic Hades. After courtesies are exchanged, to win Arawn's friendship, a strange deal is struck. They exchange places—Pwyll descends into Annwvyn to help rid Arawn of an enemy, while Arawn returns to Pwyll's castle and pretends to be him, even sleeping with his wife (while refusing her any affection). Several such inversions occur in *The Mabinogion/Y Mabinogi*—reversals of fortune that show not only a disruption of the status quo but a transference of the magical into the mundane and vice versa. Enchantment is never far away. Later, in Pwyll's tale, we hear of how he stands upon a magical hill of vision near his 'chief palace' in Narbeth: Gorsedd Arbeth, where it is said one may receive a wound or behold a wonder. Fortunately, for Pwyll, he is granted a vision of Rhiannon upon her fairy steed: 'a lady, on a pure white horse of large size, with a garment of shining gold around her, coming along the highway that led from the mound' (Guest, 2000: 6).

At first Pwyll sends a retainer to follow her and to find out her name, but no matter how fast he rode he could not catch her, though she only progressed at a 'slow and even pace'. When Pwyll pursues himself, he too cannot catch her until he calls out, 'O maiden, for the sake of him whom thou best lovest, stay for me' (2000: 8). Rhiannon stops, and reveals she had intended to seek Pwyll, because of her love for him. They agree to marry, although further complications ensue. Yet, it is telling that the correct ritual phrase must be uttered, the correct etiquette observed.

In the Scottish Borders a 13th-century ballad, 'Thomas the Rhymer', relates how the titular poet (a historical figure renowned in that area for his prophecies) is granted a similar tête-à-tête with the 'Queen of Elfland', who rides into his vision upon a fairy horse while he idles by the Eildon Tree, at the foot of the Eildon Hills. She too seems to have come seeking him, and, taking him up behind her on her steed, she rings her magical bridle, and they journey to Elfland—an initially drear landscape, sunless, moonless, and flooded with blood-red water. They come to a tree, and there the Queen shows Thomas three roads: to Heaven, Hell, and 'fair Elfland'. He chooses the latter, and there he is absconded for 'seven mortal years'. He is made to pledge not to reveal anything about what he experiences, and is given 'the tongue that cannot lie', the double-edged gift of prophecy, in exchange, along with a pair of elven boots and a cloak of green: the livery of Faerie.

When Oisín, poet of the Fianna (Fionn Mac Cumhail's famous warrior warband) and son of the great Fionn himself, meets Niamh of the Golden Hair on the western shores of Ireland, he is not so fortunate. Although he is, like Pwyll and Thomas, sought after as an eligible bachelor by a Fairy Queen, and summoned to Tir nan Og by Niamh's enticing song, there to woo and wed her, what seems like a pleasant holiday for him lasting no more than a few weeks is revealed to be a 'chrono-abyss' (Clute and Grant, 1999), for when he returns to Ireland he finds three hundred years have passed and all of those he has known and loved have turned to dust. Although warned by Niamh not to touch the earth, when Oisín bends down from his saddle to help some farmers move a rock, his straps break, and he plunges to the ground. As soon as he touches terra firma, his fairy steed disappears, and he is overwhelmed with extreme old age. Taken to a nearby monastery, it is St Patrick himself who hears Oisín's remarkable tale before he expires—a traveller out of time.

A similar fate befalls the Children of Lir, who are turned into swans by their stepmother Aoife out of jealousy. They are cursed to spend nine hundred years in their avian forms, forced to migrate every three hundred years across Ireland. Finally, a monk takes the swan-children in—linking them by a silver chain as a symbol of the strong sibling love that has protected them in their travails. The final detail of Aoife's curse is that they must remain in swan-form until a 'woman of the south marries a man of the north'. This occurs, and the children are returned to their original human form, but by now are ancient, and quickly expire—their last wish to be buried together.

Rhiannon is accompanied by otherworldly birds, whose song is said to transport listeners into Faerie; Llew Llaw Gyffes in turned into an eagle by the betrayal of his fairy bride, Blodeuwedd (made from the flowers of oak, and broom, and meadowsweet by the magicians, his uncle Gwydion and Math, son of Mathonwy). After Gwydion frees his nephew from his feathered prison, he punishes Blodeuwedd by turning her into an owl (a story Alan Garner contemporised in *The Owl Service,* 1967). The mighty giant, Bran the Blessed, was associated with ravens. And in the story of Mabon, son of Modron, 'who is stolen from between his mother and the wall when three nights old'—is advised to seek out the oldest of animals for advice—instigating an epic quest to discover his whereabouts by a company comprising Culhwch, Gwalchmai (Gawain, the Hawk of May), and others of Arthur's court. These 'oldest of animals' include a blackbird, an eagle, and an owl. Yet it is the Salmon of Llyn Llwyd he ultimately helps them hunt down Mabon—who is trapped behind the walls of Gloucester. The *puer eternus* is freed, and Culhwch can complete *his* quest—the tasks set him by Ysbaddeden to win the hand of Olwen, another 'flower-bride'. Time and time again, it is the relationship with the natural and supernatural worlds that creates harmonious equilibrium, flourishing, and the healing of the Wasteland.

2 Children's Fantasy

The Mythopoeic Landscapes of Childhood

In a survey of 1,000 children aged 7–14 years, The National Trust found that two-thirds of young people want to spend more time in nature, and yet 63% of parents also consulted said they were only able to take their children to natural places on one day a week or less mainly due to accessibility issues (*Guardian*, 2024). The National Trust director general, Hilary McGrady, said, 'Research shows that if children and young people can engage with nature early in life, they grow up to care about the natural world and are more likely to take action to protect it' (*Guardian*, 2024). And yet, earlier research indicated a worrying decline in access to green space by young people. A 2012 report into 'Natural Childhood' states that between 1970 and 2010, 'the area in which children were permitted to play unsupervised shrank by 90 per cent', and the proportion of children playing regularly in 'wild' places fell from one in two to one in ten (cited by Macfarlane, who concludes: 'The disconnection from nature is greater now than it has ever been' (2015: 324)).

Childhood can contain moments of magic that last a lifetime—that we can spend our whole lives trying to return to, or revive. Rife with anxieties about dangers real and imagined, it is a time when make-believe is a default for many—as we continually seek to vault the narrow frame of our experience with flights of fancy. Leaping the constraints of our age, gender, physical appearance, fitness, strength, capabilities, and socio-economic class, we imagine other lives, other ways of being in the world. Although each childhood is unique in some respects—and sometimes vastly different depending on where one is born in the world, into what country, culture, and geopolitical moment—there are also universal elements. And many of these have become tropes in Children's Literature: the death of one or both parents; overcoming the bully; the magical friend or pet; the unfamiliar visitor; reversals of fortune; the discovery of a hidden power. The Aarne-Thompson-Uther Index (1910,

DOI: 10.4324/9781003463665-3

1928, 1961, 2004) seeks to provide a comprehensive taxonomy of these 'tale-types':

1–299	Animal Tales
300–749	Tales of Magic
750–849	Religious Tales
850–999	Realistic Tales
1000–1199	Tales of the Stupid Ogre (Giant, Devil)
1200–1999	Anecdotes and Jokes
2000–2399	Formula Tales

What concerns us here are tales that fall within the purview of the Fantastic. At least three of these categories clearly do: Tales of Magic; Religious Tales; and Tales of the Stupid Ogre—although fantastical elements may appear in any, except perhaps Realistic Tales (although mimesis often provides a frame for the fantastic, and is not antithetical to it, as Attebery (2022: 25–43) has argued).

However useful as a survey of influential source texts, Aarne, Thompson, and Uther based their index upon folk tales, and so we need not be bound by it here, as Fantasy draws from many wells. The landscape of Fantasy is vast, unruly, and forever shifting. Yet, some landmarks coalesce and linger when generations of writers, readers, critics and fans have returned to them again and again and given them substance. These 'landmarks' are formed not just from folk tales but from the whole spectrum of human experience, including our deepest fears and strongest desires. They include The Magical Cottage; The Secret Gate; The Hidden Garden; The Stream; The Island; The Magical Child; The Wood; The Dark Stranger; The Wheel of the Year; and Animals and Other Critters. Others no doubt exist (e.g. The Strange House, often featured in School of Magic narratives such as the *Summer Magic/The Journal of Luke Kirby* series (1988–1995)), but these are evidenced by some of the greatest works of Children's Fantasy, and provide a tentative map of the fledgling imagination, akin to 'An anciente mappe of Fairyland: newly discovered and set forth: imaginary locality' (1920) by Bernard Sleigh, which depicts many of the Secondary Worlds of Fantasy Literature.

These 'landmarks' are often informed by the real mythscapes of childhood—places of exploration, and enchantment. For children, 'place' is rarely a single, static concept, but one plastic and porous with their imagination. As Macfarlane points out, this endlessly rich mythologising of the local landscapes of childhood is the loam of Children's literature: 'The children's literature understands this different order of affordance. This is why so many of the greatest children's stories involve thresholds, place-warps, time-slips and doorways: access points that lead to experience and danger, in defiance of standard geometries, and often beyond the guardianship of adults' (2015: 316).

So, in a complex ecosystem of reciprocity imagination enriches landscape, and vice versa.

For our purposes here, we shall focus on the more obviously environmental 'landmarks'—what could be called sites of initiation and transformation—and how they explore the Ecological Imaginary.

The Magical Cottage

> *'There is a place called 'heaven' where the good here unfinished is completed; and where the stories unwritten, and the hopes unfulfilled, are continued. We may laugh together yet.'*
>
> J.R.R. Tolkien (Letter to his son, 9th June 1941; cited in Carpenter, 1995: 54–55)

Tolkien ostensibly wrote *The Hobbit* for his son, Christopher, but there is a strong likelihood that he was ultimately writing it for himself—for his own 'inner child' and for the memory of his friends from the TCBS, two of whom were lost in the First World War (Garth, 2003)—for the project ultimately extended far beyond a bedtime story for his little boy, in effort and scope. This is not to say his Middle-Earth magnum opus was merely dabbling in 'childish things'—for he approached it with all the powers of his professorial mind, genius as a writer, and gifts as a philologist—but there is perhaps an element in all Fantasy writing which is a yearning to return to the 'garden of lost innocence', whether that be of childhood, early human evolution, or an Edenic pre-lapsarian state.

In writing Fantasy it could be argued that we are all trying to get back to the garden in one way or another, whether we imagine Secondary Worlds of arboreal splendour, Medieval feudalism, deep magic, or assassinatory sang-froid: it is all playing dressing up in the attic of our imaginations, or hide and seek at the bottom of the garden.

Christopher Tolkien became literary executor of his father's estate, and published posthumously a 12-volume *History of Middle Earth* (1983–1996). Amongst those, he edited a 2-volume collection of J.R.R. Tolkien's fragmentary or unpublished stories called *The Book Of Lost Tales* (2002 [1983]). One of those is 'The Cottage of Lost Play'. The basic storyline is that a traveller named Eriol comes to a fairy land by ship. He wanders to the centre of a great island and weary, looks for shelter for the night in a small cottage in the village at the centre of the island. The name of the house in Elvish is 'Mar Vanwa Tyalieva' or the Cottage of Lost Play.

Lindo and Vaire are his hosts. With them are many of their kin as well as friends and children of many ages. Eriol hears a summoning gong. Vaire calls the room to which they go the 'Hall of Play Regained' (2002: 15). It is where tales are told and is filled with children and people of all ages. Here is an extract of the poem Tolkien wrote about it:

The Little House of Lost Play (Mar Vanwa Tyaliéva)

We knew that land once, You and I,
and once we wandered there
in the long days now long gone by,
a dark child and a fair.
Was it on the paths of firelight thought
in winter cold and white,
or in the blue-spun twilit hours
of little early tucked-up beds
in drowsy summer night,
that you and I in Sleep went down
to meet each other there,
your dark hair on your white nightgown
and mine was tangled fair?

Here the liminal state necessary for entering such otherworlds is suggested in phrases laden with yearning such as 'the paths of firelight thought' or 'the blue-spun twilit hours'. These are akin to Yeats' 'when moth like stars are all alight'—when Wandering Aengus lights a fire, glimpses his 'glimmering girl', loses her and ultimately finds the binary opposites alluded to in Tolkien's 'dark hair on your white night gown/ and mine was tangled fair' in the 'silver apples of the moon, the golden apples of the sun'. It seems we need these polarities to navigate by and between, like the clashing rocks of the Symplegades. Whether they are the hemispheres of the left and right brain; the subconscious and superconscious; the Qabalistic pillars of Mercy and Severity; the Apollonian and Dionysian; or the Divine and Diabolical—between such extremes we find our doorway.

Lindsay Clarke, in his essay 'Imagining Otherwise' (2004), explores this notion through the ancient symbol of the *vesica pisces*—which he sees a 'locus of . . . collision' between the inner and outer world. He writes that 'it is through the meeting of opposites that something new appears'.

Yet the challenge is holding these polarities in balance long enough for something of wonder to slip through. Clarke articulates this creative tension: 'the proper exercise of the imagination is an energetic process of negotiation between the outer world and the inner world through which we strive to hold together their often contradictory pulls' (2004: 11). Reality is always threatening to transgress the fragile borders of Fantasy—a threat which gives such polders (to use Grant and Clute's use of the term) as Narnia, Middle-Earth, or Neverland an evanescent fragility and poignancy.

This is epitomised by Tolkien's poem about him and his brother—a lost Arcadia of childhood innocence and imagination. The poem paints a scenario so fey and whimsical it is doomed to corruption—an Eden we must be exiled from to gain Knowledge. Yet, here worlds were born.

Tolkien and his brother were not unique in forging worlds of Fantasy from the gardens of childhood. The Brontë siblings created their own world, two imaginary islands called Gondal and Gaaldine (was Tolkien's Gondor indirectly inspired by these?), which they populated with knights and damsels, emperors and princes. This was the nursery of their imaginations, and from this cutting of teeth in a mutual storyworld came the authors of *Wuthering Heights* (1847), *Jane Eyre* (1847), and *The Tenant of Wildfell Hall* (1848). It is worth looking at this Juvenalian world-building in detail for the insight it gives into fledgling fabulists.

All the Brontë children (Charlotte, Anne, Emily, and Branwell) were involved from fairly early childhood in writing the chronicles of imaginary countries in elaborate detail with great gusto. Mr Bronte's return from Leeds with a box of wooden soldiers for Branwell on 5th June 1826 first released this extraordinary imaginative energy. Each child chose a soldier and determined his name and character, and these characters became actors in a series of games that grew steadily more elaborate as more characters were invented and the actions in which they were involved grew into complicated serial stories. Written out meticulously in a neat and tiny hand, the stories form an extraordinary collection of miniature home-made manuscript books. The surviving Brontë juvenilia have been carefully examined by Fannie E. Ratchford and discussed by her in her 1941 book *The Brontës' Web of Childhood* (the title of the book comes from Charlotte's poetic account of these literary activities, 'We wove a web in childhood'). Her conclusion is that the little booklets are a closely connected series of stories, poems, novels, histories, and dramas having a common setting and common characters, written through the sixteen years between 1829 and 1845: 'They are the epic cycle of an imaginary world in which the four young Brontës lived, moved, and had their being, into which they projected themselves and all that they knew of life from narrow experience and wide reading' (1949). The surviving work is mostly Charlotte's—she and Branwell developed their cycle of Angria—while the stories of Gondal, which were developed independently by Emily and Anne, have not survived. All that we have to indicate the nature of the Gondal cycle are poems by Emily and Anne spoken by Gondalian characters at critical moments in their history, and a very few prose fragments referring to the progress of the stories or identifying a place or a character.

However, we have a precious insight into the weaving of this 'web of childhood'. On 24th November 1834, when she was sixteen years old, Emily captured a vivid vignette of creative life at Haworth parsonage:

> 'Taby said just now Come Anne pilloputate [peel a potato] Aunt has come into the kitchen just now and said Where are your feet Anne Anne answered On the floor Aunt. Papa opened the parlour door and gave Branwell a letter saying Here Branwell read this and show it to your Aunt and Charlotte. The Gondals are discovering the interior of Gaaldine. Sally Mosely is washing in the bock kitchen.'

And there we have that enticing phrase: 'The Gondals are discovering the interior of Gaaldine'. The completely matter-of-fact way in which this remark about the progress of the Gondal epic is inserted into the account of ordinary household affairs indicates the degree to which the events of this imaginary world were taken for granted by Emily as an integral part of the world she lived in. Gaaldine, as other clues indicate, was conquered and colonised from Gondal. Anne added in pencil to the list of proper names at the end of the family's copy of Goldsmith's *A Grammar of General Geography* (1800) a list of Gondalian place names which provide some important clues for the reconstruction of the stories. 'Gaaldine, a large Island newly discovered in the South Pacific' comes immediately before 'Gondal, a large Island in the North Pacific'. We know from the poems that Gondal had a climate and landscape like that of Yorkshire, while Gaaldine had a landscape of tropical luxuriance.

It is a curious fact that Tolkien's own mythos alludes to a similar Atlantean 'lost island', Numenor, and he was haunted all his life by a recurring dream which filled him with a dread of inundation. He puts this into the mouth of Faramir, in *The Return of the King* (1955), who describes his vision of a 'great dark wave climbing over the green lands and above the hills, and coming on, darkness unescapable. I often dream of it'.

Primal fears and desires seem to be the pistons and petrol in the engine of the imagination.

David Daiches, in his introduction to *Wuthering Heights*, observes, 'The most impressive and passionate of Emily's poems are Gondal poems, and even though they obviously reflect her own deepest emotional needs they are on the surface dramatic utterances by characters. Some of these characters, and some of the situations in which they find themselves, remind us vividly of characters and situations in *Wuthering Heights*. . . . Emily Brontë's strange and powerful novel thus sprang from a lonely but well-stocked imagination' (1965).

Many other writers were inspired by formative experiences, from childhood games of make-believe and the personal cosmology of their neighbourhood—the world on their doorstep, and what lay beyond. We all start off with an imperfect knowledge of our vicinity, and how the world works—and fill in the gaps.

Alain-Fournier (the pen name of Henri-Alban Fournier), a French author killed in action in 1914, wrote a single classic novel, *Le Grand Meaulnes* (1913), whose title, although referring to the daring young protagonist, can be translated as 'The Lost Domain'. The novel is a paean to the magic of childhood, when the borders of one's world were still being probed, crossed, and regained. The schoolboy hero 'borrows' a powerful horse and carriage, in a Phaeton-like act of bravado and folly. He goes missing for three days and returns with an intriguing dreamlike account of an adventure into a 'mysterious domain', which he strives to return to for years—'in search of the lost trail' that will lead him to his sweetheart, a quest that leads to a bitter awareness: 'I'm sure now that when I discovered the nameless domain I was at some peak of perfection, of purity, to which I shall never again attain. Only in death, as I once wrote you, can I expect to recapture the beauty of that moment . . .' (1913: 148).

In his memoir of reading, *The Child That Books Built* (2002), Francis Spufford explores the evolution of his imagination through the childhood classics which influenced his development. In describing his experience of reading a book, he says, 'You never came out the way you came in.' Each talismanic book was a rabbit hole of possibility, of unexpected transgression and transformation.

Fantasy author Robert Holdstock suggests the conversations and behaviour of adults eavesdropped upon by children create a psychic landscape of larger-than-life characters, heroes, and villains—what Carl Jung termed 'archetypes', the psychoanalyst's phrase for 'primordial images': conscious representations of eternal principles; representations that can vary a great deal without losing their basic pattern. 'They are without known origin and they reproduce themselves in any time or in any part of the world' (Jung, 1964: 58). For Holdstock, the stuff that archetypes are made of—mythic images—inspired his neologism 'mythagos'. These emerge from the symbiotic interaction between his characters and a remnant of ancient woodland, Ryhope Wood, in his 'Mythago Wood' cycle of novels. It is suggested the 'mythagos' are plucked from a deep and ancient part of the protagonist's brain—a mysterious process which seems a mirror of Holdstock's own creative process: 'All things are known, but most things are forgotten. It takes a special magic to remember them' (1988).

The gateway of Ryhope Wood (a classic example of 'Threshold Fantasy'—what Farah Mendelsohn might call 'Portal Quest' or possibly 'Liminal', for it has elements of both) (2008) seems to be an amalgamation of an actual remembered landscape from childhood, and imagination, as Holdstock describes:

> 'There was a Roman villa on the hillside behind my house. I have no idea how old I was—four perhaps—when I began fantasising about the Roman ships coming up to the harbour there. That's where Romney Marsh is now, but in Roman times it was under the sea.'
>
> (cited in *Wordsmiths of Wonder*, 1993: 100)

Thus, Holdstock's storyworld was formed from his own 'lost island'—the reclaimed land of Romney Marsh in Kent, 'reclaimed' by his imagination, that might one day be reclaimed by the sea again if Global Warming persists. And similarly our own precious outliers of imagination (suffering the islandisation of 'mature' materialism, which insists such toys be put back in the box so that we can get on with the serious business of living) are threatened with engulfment and oblivion—by the tides of mundanity and commerce, or the marine transgressions of the mainstream.

The storytellers told themselves stories before telling the world. They began by desiring dragons—that is, creating Fantasy worlds—and we became infected with their feverish enthusiasm, because the longing is inside all of us (Manwaring, 2014). Later, it may be channelled into something more practical or at least attainable (motorbikes; mountains; obscure records or books), but for a time-being we are all united in the garden of lost play: 'In one thing only were all alike, that a look of great happiness lit with a merry expectation of further mirth and joy lay on every face' (Tolkien, 2002: 15).

This quality is perhaps the secret drug we look for in all Fantasy writing—although it might be buried beneath a myriad of masks, sub-genres, styles, and tones. It is the *hiraeth*—the Welsh term for 'deep longing for something, especially one's home' (Oxford Languages)—for this which makes us return again and again to the shelves, looking for the book that will act as a portal to the place we once experienced (or wished we had) and yearn to return to, forever exiles in our sensible adult world.

The Hidden Garden

Before such a place can be entered, the protagonist/s of many Children's Fantasy have to discover and gain entry to some kind of portal. In the Campbellian Hero's Journey, this is the threshold that separates the Ordinary World from the Special World: transition through it marks an ontological shift as the characters leave all that is familiar and enter a disorientating new realm whose rules are as yet unlearnt, and whose inhabitants could be friend, or foe, or something in-between. This resonates so greatly with young readers partly because they themselves will experience such psychically painful exits and entrances throughout their infancy and youth—from the existential and sometimes very real trauma of birth, being weaned from the breast, the harrowing first day of school and the transition to older schools, moving house, experiencing heartbreak and death for the first time, and, eventually, puberty. As a trope, some kind of portal features prominently in classics of Children's Fantasy: the rabbit hole in *Alice's Adventures in Wonderland* (1864), the forest of pools in *The Magician's Nephew* (1955), and the wardrobe in *The Lion, The Witch, and the Wardrobe* (1950), the titular tollbooth in *The Phantom Tollbooth* (1961), and the roadworks in *The Subtle Knife* (1997). One of the most famous and enticing portals in all of Children's Literature is the locked door that leads to The Secret Garden. Frances Hodgson Burnett's much-loved classic (published in toto in 1911 after being previously serialised in *The American Magazine*) is not a Fantasy narrative in the strictest sense—in terms of containing elements considered impossible in a materialist reality—being predominantly 'rationalist-Gothic-inflected' (Levy and Mendlesohn, 2016: 47); yet it reifies key aspects of Portal Quest: the protagonist transitions from an 'ordinary world' to a 'special world', assisted by a guide (or, in this case, several—the human and more-than-human) via the trope of a difficult-to-access ontological smeuse. Although the titular garden in Hodgson Burnett's novel is not a Secondary World, it comprises aesthetic affinities with the heightened register, immanence, and enchanted nature qualities of many classic storyworlds. Although initially in a fallen, or neglected state, the garden eventually brims with a paradisal, prelapsarian fecundity. The 'lost' state of this walled garden, and its restoration, instantiates a personal 'paradise found'—a healing return to enchantment after trauma. The inciting incident of this narrative is a catastrophic Cholera epidemic (depicted as an earthquake in the 1993 film adaptation), which kills off the urbane socialite parents of Mary Lennox, who had been enjoying a well-heeled colonial existence in mid-19th-century India. Repatriated along with other survivors of the disaster, Mary is made the ward of a reclusive uncle, lord of a sizeable, but unwelcoming estate in the north of England, Misselthwaite. The widowed, gloomy, and hunchbacked Archibald Craven cuts a Byronic figure—with his spinal deformity a symbolic echo of the Romantic poet's clubfoot—but is telling by his absence in the early stages of the book. Instead, Mary is put under the charge of the redoubtable Mrs Medlock, who runs the house with a firm hand in the absence of its patriarch. The 'spoilt' Mary is initially insufferable in her manners, behaviour, and expectations of servitude—although she cannot be blamed for the colonial circumstances of her upbringing, and she is clearly traumatised from being orphaned at such an early age. Frozen in her grief, and unable to cry, the narrative arc describes the emotional thawing that occurs in parallel with the garden's restoration—a garden created for the much-beloved late Lady Craven, and kept under lock-and-key ever since, until Mary discovers the garden one day guided by a friendly robin, and then the key in Lady Craven's former bedchamber. This neglected 'secret garden' is symbolic of the cauterised emotional life of the Lord, and is echoed in the garden in Oscar Wilde's fairy story, *The Selfish Giant*; the rewilded wings and owl-inhabited Tower of Flints in *Gormenghast*; the cobwebbed rooms of Mrs Haversham in *Great Expectations*; and the labyrinthine, flooded ruins of *Piranesi,* to name a few examples. Throughout Hodgson Burnett's novel, nature is depicted as healing, redemptive, and non-hierarchical. The personification of this more healthful, organic, umwelt is Dickon—a young, working-class boy whom Mary befriends in the grounds of Misselthwaite. Able to communicate with animals, many of whom are his constant companions, Dickon is a Pan or Puck figure—a *puer aeternus*—one who haunts late 19th-century and early 20th-century literature, for example, *The Great God Pan* (Machen, 1890); *Peter Pan* (Barrie, 1902–1911); 'The Man Who Went Too Far' (Benson, 1904); *The Wind in the Willows* (Grahame, 1908); *Puck of Pook's Hill* (Kipling, 1906); 'The Touch of Pan' (1917), *The Blessing of Pan* (Dunsany, 1927); *The Crock of Gold* (Stephens, 1928)—as well as art, music, and theatre, a cultural phenomenon historian Ronald Hutton has carefully examined (1999: 43–51). Dickon is the distant cousin of Enkidu. Here, he rides an unsaddled horse across the open moors outside of the control of Misselthwaite (itself, a faint Uruk/Heorot/Camelot analogue). Alongside teaching her about nature, Dickon helps Mary to reconnect with the world of play, and her own cauterised feelings and embodied self. As a symbol of 'High Wild Magic' (2016: 42), Levy and Mendlesohn suggest Dickon's role is not to guide them 'into the wild', like Aragon does the hobbits in LOTR, but 'to bring the fantastical space into the hidden garden' (2016). And this quality of containment is intrinsic to Children's Writing—the safety created by domestic settings, microcosmic narratives where there is a 'love of the miniature' (2016: 39)—most concretely in Portal Quest narratives such as *The Lion, the Witch, and the Wardrobe* (Lewis, 1950), where 'the fantastic is *contained,* and the containment itself' (2016: 39). Yet, in Hodgson Burnett's novel, the fantastic (as symbolised by the secret garden) is denied *because* of literal containment: incarceration. For, in stark contrast to Dickon, the 'sickly' heir to Misselthwaite, Colin, is kept bedbound in a darkened room with the windows shut. Archibald overcompensates for the death of his wife by smothering his son in a paradoxically debilitating system of care. Colin personifies stifled nature, but also a different kind of intelligence and way of being in the world to Dickon. Having received the benefits of a good (if isolated) education, Colin is cultivated and sophisticated, with ambitions to become a scientist.

And yet his enfeebled state cut-off from the vital pulse of life—and the embryonic eternal triangle that these three characters present (Mary with her yearning for re-connection to her natural, emotional self; the 'wild boy' Dickon' attuned to nature; and infirm Colin, who is pushed around in a Bath-chair)—seems to foreshadow the dynamic of *Lady Chatterley's Lover,* although the latter is obviously more critical of class and intensely sexualised. However different these texts may seem—the former primly Edwardian, the latter, wildly Modernist, in both, 'the garden extends the possibilities of the imagination and of reach' (Levy and Mendlesohn, 2016: 42). It offers a wild, dreaming place—ludic and liberated from the stifling conventions of society. Nature enables the imagination, encouraging us to think outside of the box, rhizomatically, and non-anthropocentrically. For many young (and adult) readers of *The Secret Garden,* it is hard to remain unaffected by the enchantment it creates—for the nature on our doorstep; or the powerful hiraeth it evokes for a prelapsarian 'land of lost content', the search for which may become a major line of desire or even life's work (Nanson, 2023).

Now, let us leave the garden and follow the stream—flowing forward a generation into the shadows of the Second World War, where we find unexpected magic.

The Stream

The Little Grey Men (1942) is a Children's Fantasy classic by D.J. Watkins-Pitchford, written under the pen name of 'BB'. Published in the darkest years of the Second World War, it won the Carnegie Medal. It relates the adventures of the 'last gnomes of England'—Dodder, Baldmoney, and Sneezewort, who go in search of their erstwhile companion, Cloudberry, who has gone missing while seeking the source of Folly Brook like a diminutive Percy Fawcett. Hidden away in a peaceful corner of Warwickshire, the gnomes dwell under the roots of an old oak ('Oak Tree House') downstream, and co-exist peacefully with most of the fauna along the small river (the 'Stream People'), with the exception of 'Wood Dogs' (foxes), stoats, and a trigger-happy farmer ('Giant Grum'), who blasts anything in Crow Wood except his pheasants. The Ecological Imaginary is vividly foregrounded in this novel. Not only is there a tangible sense of deep place—as in *Puck of Pook's Hill,* but a vertiginously ancient sense of place is also evoked via the gnomes, who have preceded most life on land, it seems. Sneezewort was considered the youngest of the three friends, despite being at least two millennia old: 'he was born long before Julius Caesar landed in Britain, but that is awfully young for a gnome' (BB, 2012 [1942]: 217)—but the author, who was an amateur naturalist, saturates the novel in well-observed riparian texture, clearly the result of spending many happy hours messing about on or near water. BB writes about nature not in a generic way, but in a microcosmically specific way. The flora and fauna, and affect, of his chosen locale, is evoked in loving detail. So much so that it has inspired generations of readers to seek similar delights. The nature-writer Melissa Harrison (who has gone on to author her own children's books in 'tribute' to BB's gnome books) cites it as one of her favourite childhood reads. She was impressed by how it didn't write down to the young readership, but, by carefully observed (and realistic) detail about the ecology of Warwickshire, 'trusts them to build up a detailed imaginative picture of a landscape through the seasons.' And yet, this vision, however situated, is still frustratingly out of reach, as Harrison opines: 'Due in no small part to The Little Grey Men and much of the rest of our childhood library, I grew up in love with a vision of England I couldn't find anywhere, and which may never have existed' (2021). Nevertheless, Harrison's imagination was stirred, and her future career nurtured by her formative reading. Yet the foregrounding of the natural world, while an undoubted strength and enduring quality of BB's novel (when other aspects—such as the absence of female characters, and apparent xenophobia: rabbits, grey squirrels, and pheasants are spoken of disparagingly for being non-native) could have completely subsumed the fantastic—with the gnomes providing a child-friendly device for simply celebrating English rivers—if not for the later chapters. In 'The Gathering of the Wood People', BB inserts his own equivalent to 'The Piper at the Gates of Dawn' chapter from *The Wind in the Willows*—after the shooting of his friend, Otter, Dodder decides to summon the more-than-human dwellers of Crow Wood to deal with their menacing tyrant resident human, 'Giant Grum'. Failing to find a way to communicate with them all in time for the auspicious Midsummer Night (June 24th)—a day hallowed in English Literature since Shakespeare's Comedy, but worshipped in Britain for at least five thousand years, as Stonehenge's midsummer alignment attests—the elder gnome prays to Pan, who magically appears before the gathered Wood People as a kind of *deus ex sylva* (2012: 133–144). And, later on, further upstream and later in the year, we hear of the wider supernatural ecology of Britain, when Dodder relates a story of his encounter with the 'Heaven Hounds' (snow geese), 'many cuckoo summers ago': 'Quite a number of gnomes lived up the Folly in those days, and the country was full of them, Elves, Goblins and Hobgoblins, Pixies, Nixies, Sprites, Brownies, and Jack-o'-Lanterns or the Lantern Men.' Drawing upon the Fairy Tradition of the British Isles, BB's gnome-narrator tells of the mischief many of these supernatural critters used to make; and the meeting with the Heaven Hounds (and their invitation to join them on their migration back to Spitzbergen in the Arctic Circle) fills his brother, Cloudberry, with the *hiraeth* traditionally experienced by those who have heard fairy music, or tasted of fairy fruit, in the many tales and ballads of the tradition. The 'Land of the Northern Lights' Cloudberry later travels to is alluded to like an extant fairyland: 'I've seen Spitzbergen and the Land of the Northern Lights, and it will take the rest of the winter to tell you of all my adventures and the strange things I've seen.' (2012: 244). It is tempting to speculate that Philip Pullman took up the gauntlet to describe these 'strange things' in his own Carnegie Medal-winning book, *Northern Lights* (1995). However, even considering the co-existence of demi-gods and nature elementals, the novel remains grounded in a real sense of place, but one that is enchanted by not just these elements, but by the quality of attention the author gives to the richness discovered under our noses (which the human characters frequently miss). As with other riparian children's novels (*The Water-Babies*, *The Wind in the Willows*, *Swallows and Amazons*, *Puck of Pook's Hill*), it is the watery setting itself that becomes enchanted by association with such flights of fancy—it provides a dreaming place, as well as place of adventure, arrivals, losses, discoveries, and departures. Yet in none other than *The Little Grey Men* is the river described with such affectionate exactitude: white hawthorn petals, caught in whirlpools, revolve 'like fairy wheels' (2012: 75); a smeuse between curling fronds of new bracken is a 'fairy avenue of green' (2012: 106). Its light, shadows, motion, and sound are constantly described—the ever-renewing news from upstream. At one point, the weary gnomes fall asleep to its familiar song: 'the music of the stream is a comforting lullaby, one that was familiar, and one that had always been part of their sleeping and waking life' (2012: 89). And it is music that heralds the arrival of Pan: ' it was pipe music, surely the loveliest music ever heard. It seemed not of this earth: yet in it was all the songs of the birds, slender reeds, the song of insects at the heat of noon, of trees in tumult, the voices of secret streams and broad rivers and of the eternal seas' (2012: 135). Yet, despite such a numinous manifestation, the gnomes still bewail, 'there is no magic left in England now!' (2012: 139). However, it must have seemed so, in a starkly existential way, in the depths of the Second World War. The novel could be read as a fairy tale retelling of that conflict: home-loving Brits forced out of their comfort zone to come to the aid of their fellows. The recovery of Cloudberry could be interpreted as a loose re-imagining of the repatriation of British soldiers at Dunkirk—the improvised boats fashioned by the gnomes to traverse the waterway like the merchant fishing vessels used to bring Tommies home. The overcoming of the brutal Giant Grum by the alliance of the Wood People could be a (hoped-for) outcome of *that* conflict. The seemingly miraculous arrival of the Jeanie Deans mechanical boat, complete with precious food supplies, a metaphor for the life-saving 'North Atlantic Bus' that kept Britain going. And the victory of the humble, 'common folk', caught up in a conflict far bigger than theirs—one in which they are compelled to dig deep and find their real mettle—seems to foreshadow Tolkien's tale of courageous hobbits (in LOTR, published twelve years later, if not in *The Hobbit,* four years prior, which has the tone of a comforting folk or wonder tale, rather than a perilous epic). There is, then, a darker subtext to the ostensibly light and humorous tale of BB's—the dangerous world of humans is distant, but ever present (a constant threat to the critically endangered species of the gnomes, and brought shockingly into frame by the 'gallows' of Crow Wood, where Giant Grum has hung many of his kills); and loneliness, sorrow, and hardship is often in the wings. And, finally, however apparently conservative the trappings of this evocation of an England 'which may never have existed', with its sense of feudal hierarchy (as epitomised by the glorifying manner in which the fox hunt is described), and sleepy backwaters, BB has the gnomes cry out indignantly: 'There's no such thing as private property in nature!' (2012: 107), which could have come from the Right to Roam group, currently campaigning for better access to the countryside across England and Wales (where only 8% is accessible to the public). In such details, *The Little Grey Men* continues to live beyond its historical context, finding resonance in new generations of readers, some introduced, like Melissa Harrison, by their parents and grandparents to a much-loved classic. In BB's book, it is only children who can (still) spot gnomes—as the author himself claimed to have done so at the age of four—but it is often adult readers who advocate the book to their young. In the introduction, the author suggests it is an act of perception, of being in nature, that allows us *all* to access this enchanted realm once more: 'If you don't believe in the Little People, I would ask you to make yourself as small as possible (which is horribly difficult) and keep very quiet (which is more difficult still) and watch and wait by the streams in the woods, as I have done. And suddenly you will understand that the birds and wild animals *are* the Little People!' There is no clearer exhortation in a fantasy book—for children or adults—to go out into the natural world and experience it directly. Thus, *The Little Grey Men* remains a prime example of the Ecological Imaginary.

The Island

Islands feature prominently in classic Children's Literature such as *Treasure Island* (1883); *Swallows and Amazons* (1930–1947); The Famous Five series, which started with *Five on a Treasure Island* (1942); *and The Lord of the Flies* (1954). Therefore, it is unsurprising that islands of all shapes and sizes appear in Children's Fantasy as a key setting.

Perhaps the most famous in Children's Literature is Neverland, which J.M. Barrie suggests is 'a map of a child's mind':

> 'Neverland is always more or less an island, with astonishing splashes of colour here and there, and coral reefs and rakish-looking craft in the offing, and savages and lonely lairs, and gnomes who are most tailors,

and caves through which a river runs, and princes with six elder brothers, and a hut fast going to decay, and one very small old lady with a hooked nose.'

((1995 [1911]: 6):)

Barrie adds to this a secondary layer of mundane things, suggesting a palimpsest of the magical and the mundane ('another map showing through', Barrie, 1995 [1911]: 7).

His crowded description of a place where 'nothing will stand still' suggests that Neverland is not just a setting for story, but a *source* for stories. A child's imagination is voracious and seemingly limitless, and yet an island gives it manageable parameters and echoes the ontological status of the child—a small but expanding consciousness in a vast sea of existence: 'Of all delectable islands the Neverland is the snuggest and most compact; not large and sprawly, you know, with tedious distances between one adventure and another, but nicely crammed' (p. 7). Any peril is contained, although this threatens to spill over into nightmares in the 'two minutes before you go to sleep' (Barrie, 1995 [1911]). Fortunately, there are night lights.

The island is a setting with rich metaphoric strata—it is a place of refuge and respite from the 'real' (adult) world where normal rules are suspended. These could be temporal, as in Tir nan Og, the Land of the Ever Young in Irish mythology, where time runs faster than in the human world—as Oisín, poet of the Fianna, discovered, when he journeyed there with Niamh of the Golden Hair: when he returned to his homeland, he discovered three hundred years had passed. A similar fate awaited the Company of the Blessed Head in *Y Mabinogi*—returning from Ireland with the head of their fallen chief, Brân, they sojourn on a small island ('Gwales'—possibly Grassholm, off the Dyfed coast) where Brân's head comes to life and regales them with tales while they feasted—the trauma of their recent tragic conflict forgotten, while a forbidden door remains closed at least. When one of their company opens it out of curiosity, time and trauma flood back in, and they discover eighty years have passed. Spatial rules may also be suspended, with the island having a disorientating or fathomless quality. This may be relativistic and connected to the size of the protagonists—so the heronry in *The Little Grey Men* (1942), where the three gnomes are marooned, is scaled up accordingly. Its parameters may be satisfyingly defined, if mapped, like Never Never Island (*Peter Pan*, 1912); or mythically vague like Saint Brendan's Island (*The Water-Babies*, 1863). An aura of mystery, danger, and adventure often pervades the more mimetic islands of Children's Literature—zones of narrative compression, they provide a contained testing ground for young protagonists' capabilities. Yet in Children's Fantasy, a numinous atmosphere is often foregrounded. The prime example of this is 'The Piper at the Gates of Dawn' chapter from *The Wind in the Willows* (1908). Little Portly, a baby otter, has been missing for some days. Concerned, Mole and Rat set off in a boat upstream in the dark, knowing the moon was due to rise soon. As they travel in a dream-like fashion up the night river, their senses are sharpened to the nocturnal denizens of the river: 'the night was full of small noises, song and chatter and rustling, telling of the busy little population who were up and about, plying their trades and vocations through the night' (1980 [1908]: 129).

When the full moon rises it creates a heightened atmosphere as its silvery light transforms the riparian nightscape: 'a difference that was tremendous' (1980 [1908]: 131). As the moon sets just before dawn, the first birdsong focuses their senses on sound. As Mole rowed, Rat suddenly alert, 'listened with passionate intentness'. Only he, at first, can hear a faint music, which rises and fades, leaving him with a sense of hiraeth, which he attempts to describe to his friend: 'So beautiful and strange and new! Since it was to end so soon, I almost wish I had never heard it. For it has roused in me a longing that is pain, and nothing seems worthwhile but just to hear that sound once more and go on listening to it for ever' (1980 [1908]: 132). Fortunately, it returns—'the thin, clear happy call of the distant piping'—and calls them to a small island. Mole hears nothing until they draw near, 'but the wind playing in the reeds and rushes and osiers' (1980 [1908]). As they approach the island, the dawning light reveals the meadows, which 'seemed that morning of a freshness and a greenness unsurpassable. Never had they noticed the roses so vivid, the willow-herb so riotous, the meadow-sweet so odorous and pervading' (1980 [1908]: 133). The natural world is transformed by the otherworldly music of the Fantastic—or rather, it enhances the perception of the protagonists (and, vicariously, the readers) to its actuality.

In deep reverence, Mole and Rat make landfall on the small island and approach the source of the sound as though pilgrims entering a long-sought temple-sanctuary. At one point Rat whispers: 'This is the place of my song-dream, the place the music played to me' (1980 [1908]: 134). Experiencing a deep 'awe' that turned their limbs to water and filled them with the terror of the Sublime, the two approach the centre of the island. There, in 'that utter clearness of the imminent dawn' (1980 [1908]: 135), they behold Pan, the 'Friend and Helper' (1980 [1908]) in all his glory. They genuflect before him and 'did worship', then the sun breached the horizon, dazzling them with its light, and the 'Vision' vanished (1980 [1908]: 136). The 'kindly demigod' bestows upon his worshippers the 'gift of forgetfulness' (1980 [1908]). Their eco-epiphany wiped from their consciousness, if not embodied, memory, they decide that the island was 'the right sort of place and that here if anywhere we should find him' (1980 [1908]: 137). Confirming this, they find Little Portly, and set off home. Yet their experience haunts them like the evanescence of a 'beautiful dream', fading upon waking. As they let the boat drift downstream, the friends feel drowsy but snatches of the 'far-away music' comes back to them, and Rat recites the lyrics that arise in his consciousness: 'Lest the awe should dwell—And turn your frolic to fret—You shall look on my power at the helping hour—But then you shall forget!' (1980 [1908]: 141). Gifted with amnesia, they return to the bucolic adventures of the main narrative and the heightened register and religiosity (a nascent neo-animism) of this chapter is left behind, but its effect, like Pan's music, lingers throughout the book and has cast a long cultural shadow since—most famously in the debut Pink Floyd album, *The Piper at the Gates of Dawn* (1967).

Another example of the numinous island is in *La Belle Sauvage* (2017). Set in the six months leading up to the arrival of Lyra Belacqua to Jordan College, the plot is dominated by a boat journey down a flooded Thames ('Everything's changed after the flood' p. 451), which allows the accidental hero, Malcolm Polstead, and a 15-year-old Alice, the feisty daughter of the landlord and lady of The Trout Inn, to escape with the baby Lyra, pursued by Gerard Bonneville, an agent of the Magisterium. At one point they take refuge on a magical island, 'The Enchanted Island' (Pullman, 2017: 436–457), a place of trickery and testing. At first, the island appears inviting—warm and fragrant and bathed in moonlight, but there are unseasonal leaves on the trees, and strange lights, and songbirds, and giant blue butterflies. Upon encountering a beautiful fairy Diania, Alice quickly realises their real names could put them in peril. Nevertheless, they partake of the deliciously down-to-earth food and drink she offers—coffee, toast, and eggs. And Diania herself breastfeeds the infant Lyra, and claims her for her own. Malcolm and Alice have to win Lyra back through trickery. Malcolm discovers an alethiometer in Bonneville's rucksack, and manages to swap it for a stone inside its protective wooden box, whose complicated mechanism only he is able to open with his mechanic's mind (2017: 442–443). When Diania fails the guessing game (due to their false names), she is inconsolable, but Malcolm leaves the box for her, and they depart.

Further downriver the young travellers find themselves on another island, which seems to exist between the alternative England of Malcolm and Alice, and the world of the Fey. The passing reference to 'The Secret Commonwealth' clearly frames this as a fairy place, alluding to the monograph by the Reverend Robert Kirk, *The Secret Commonwealth of Elves, Fauns, and Fairies* (1692/1815). Here, the secret commonwealth suggests a parallel or wainscot reality where supranatural beings co-exist. The children stumble upon what appears to be an elegant garden party taking place on the river island in the grounds of a large house (evocative of

Babbington, the Oxfordshire home of the society beauty and hostess, Lady Ottoline Morrell, who hosted famous gatherings featuring the Bloomsbury set). The world and its woes—including their pursuer, Bonneville—seem to be held at bay by a wall of mist. Yet, it is clear they cannot remain there long, charged with the safe delivery of the infant Lyra to her father in London. Later, when their canoe becomes trapped by ancient iron gates, they meet a benevolent river giant, who explains about the garden party: 'That's the place where people go when they forget' (p. 491). While there as accidental trespassers, mending their damaged canoe, Malcolm had reflected 'that one day he'd come here by right, and be made welcome, and stroll among these gardens with happy companions and feel at ease with life and death' (p. 483). This situates the island of the garden party as an English Elysium, and yet when he climbs a tree to acquire some resin to help mend his boat (foreshadowing Mary Malone's discovery of amber, which allows her to see the 'dust' forbidden by the Magisterium in the final book of *His Dark Materials*), Malcolm is granted a stark reminder of the far from paradisal 'fallen' state of the environment—an awareness granted him by this 'tree of knowledge': 'beyond it he saw a desolation, a wilderness of broken buildings, burned houses, heaps of rubble, crude shanties made of shattered plywood and tar-paper, coils of rusty barbed wire, puddles of filthy water whose surfaces gleamed with the toxic shimmer of chemical waste, where children with sores on their arms and legs were throwing stones at a dog tied to a post' (p. 483). Pullman seems to be implying we have exiled ourselves from the garden by our mistreatment of nature and the more-than-human.

An intrinsic aspect of an island is its finitude. If one wishes to continue one's journey of life—return to the adult world in the case of the young protagonists of Children's Fantasy—then there comes a time when one must leave, or be banished. Childhood is the Eden we are all cast out of by the biological fact of puberty, adolescence, and ageing. This is emphasised hauntingly in *Peter Pan,* when J.M. Barrie discusses the differing Neverlands of his human child protagonists: 'On these magic shores children at play are for ever beaching their coracles. We too have been there; we can still hear the sound of the surf, though we shall land no more' (1995: 7).

The prelapsarian quality of these riparian paradises is emphasised by the often rich and *unsullied* biodiversity they contain—so many Galapagos Islands destined to exile us from a Theocentric universe by the evidence of natural selection, which culminated in the Darwin/Wallace Theory of Evolution. The bitter paradox of paradise is we often despoil it with our presence through the 'contamination' we bring—of our mortality and morality, or more prosaically, pollution, disease, invasive species, and colonialist exploitation. The acknowledgement of this shadow is best dramatised in *The Book of Earthsea* (2018), which drew upon Jungian concepts of the 'Shadow', as well as Taoist philosophy, and the anthropology of Ursula K. Le Guin's father, Alfred Louis Kroeber. Le Guin establishes her Secondary World archipelago in *A Wizard of Earthsea* (1968), although it was foreshadowed by two short stories: 'The Word of Unbinding' (1964) and 'The Rule of Names' (1964). In the opening chapter, set on the island of Gont with its one giant peak, the archipelago is established as a place of adventure and opportunity: 'many a Gontishman has gone forth to serve the Lords of the Archipelago in their cities as wizard or mage, or, looking for adventure, to wander working magic from isle to isle of all Earthsea' (2018: 7). Through the six books of *The Book of Earthsea* Le Guin takes us to many, but not all (in refutation of Wynne Jones' 'no Tour is completed without a map. Further, you must not expect to be let off from visiting every damn' place shown on it' (2004: 131)) of the islands, most notably Roke—home of the wizards of Earthsea. In 'A Description of Earthsea' (Le Guin, 2018: 899–920), Le Guin outlines the key aspects of her Secondary World: its peoples, politics, geography, magic, and dragons, and one can see evidence of her embryonic world-building in the hand-drawn maps and notes displayed in 'Realms of Imagination' (Kirk and Sangster, 2023), and discussed in detail by Brian Attebery in the accompanying book. In creating Earthsea, Le Guin drew upon the tales of Homer told to her by her father—and one can see traces of *The Odyssey* in her protagonist, Ged's, peregrinations, as well as vestiges of the echtrai of Bran and St Brendan (Manwaring, 2008). The Irish literary tradition of the echtrai or immrama is taken to its extreme in *The Little Prince* (1945), Antoine de Saint-Exupéry's charming tale of the otherworldly protagonist, hailing from Asteroid B-61, 'a planet no bigger than a house', who travels from micro-planet to micro-planet in search of meaning. He encounters a king, a clown, and businessman, and so on—each one trapped in their particular paradigm, unlike the titular prince, who like Saint-Exupéry (an aviator) is able to travel between different modalities, and to see the world from a different perspective as he describes in his travelogues, *Courrier Sud/Southern Mail* (1929), *Vol du Nuit/Night Flight* (1931), and *Terres des Hommes/Wind, Sand, and Stars* (1939). Joining the Free French Air Force in 1943, Saint-Exupéry went missing in a reconnaissance flight over the Mediterranean on 31 July 1944. Wreckage was found off the coast of Marseille in 2000, but his body was never recovered. The tragedy of his loss bestows upon his writing, especially The Little Prince, which was published posthumously, an evanescence and poignancy, metonymic of the lost generations of all conflict.

Now let us consider the magical child in more detail.

The Magical Child

The magical child is a common trope in Children's Fantasy. Although an aspect of this—destinarianism—has been critiqued as a cliché that promotes a kind of exceptionalism (Levy and Mendlesohn, 2016: 133), especially in Post-Potter Fantasy where it has become the dominant narrative device, there is something redeemable in this figure. To Jungians, he represents the *puer aeternus* (Latin: eternal boy). Depicted in Classical art and featuring in Ovid's Metamorphosis as the minor child-god Iacchus, an integral part of the Eleusinian Mysteries, the *puer aeternus* seems to represent a perpetually renewing regenerative principle—a junior iteration of the Green Man, and the personification of Dylan Thomas' 'the force that through the green fuse drives the flower'. As such, he is the embodiment of the Ecological Imaginary. He is often depicted being at one with nature—competent, capable, and able to converse with the more-than-human: Richard Jefferies' *Bevis: the Story of a Boy* (1882), Mark Twain's *Huckleberry Finn* (1884), Rudyard Kipling's *Puck of Pook's Hill* (1906), Frances Hodgson Burnett's Dickon (1911), J.M. Barrie's *Peter Pan* (1911), Antoine de Saint-Exupéry's *The Little Prince* (1945), and Clive King's *Stig of the Dump* (1963), to name a few, the *puer eternus* is, in a narratological sense, Enkidu's younger brother. And yet there are an increasing number of female variants, for example, Alan Garner's Susan (1963), Tallis from Robert Holdstock's *Lavondyss* (1988), Ursula K. Le Guin's Therru/Tehanu (1990), Hayao Miyazaki's *Princess Mononoke* (1997), Philip Pullman's Lyra Belacqua (1995-), Nnedi Okorafor's Onyesonwu (2010), and N.K. Jemisin's Damaya (2016), to name a few. Such characters are especially attuned to nature with heightened senses and a deep connection to their environment—a sometimes self-appointed guardian of place. The righteous indignation, exasperation at the hopelessness or disinterest of adults, frustration at lack of influence, despair in the face of environmental destruction, and anarchic transgression of societal boundaries seem familiar now—and embodied in the real world by the Swedish eco-campaigner, Greta Thunberg, who started her weekly Climate Strikes as a 15-year-old schoolgirl and inspired a global movement. Along with her contemporaries—Xiuhtezcatl Martinez, Nyombi Morris, Licypriya Kangujam, Luisa Neubauer, and many others—Greta Thunberg is perhaps the antithesis of the magical child: she is real, articulate, vulnerable, and wanting action *now.* One can argue that there is *too much* focus on pandering to the *puer eternus* in modern society—with the infantilising strictures of the 'nanny state' and the immaturity of much popular culture, focused on the arrested development and spending power of the 'adultescent'. And yet there will always be a place for stories—to inspire, motivate, console, and enlighten. The magical child is a resilient figure—archetypal and perennial—they will always return, especially when things are at their grimmest and most mundane, to remind us of the immanence of enchantment in the everyday and the wonder of the natural world.

The Wood

Woods have been perhaps the primary setting for not only Children's Fantasy, but for storytelling in general for as long as stories have been told (Mellon, 1988; Maitland, 2012; Parker, 2020). In a wood, trees obscure sightlines. Things—bandits, critters, spirits—can be obscured. Roots trip, brambles snag, berries poison. Strange mushrooms appear overnight. Weird noises can be heard in the dark. The more-than-human is abroad—scenting, listening, and watching our every clumsy move. Evolved to traverse the savannah, humans may feel ill at ease in a sylvan environment. Ungainly. It is easy to become disorientated. Lost, like so many *Babes in the Wood*. This much-adapted story was first published as a broadside ballad in 1595, and later reworked by Richard Barham in his *Ingoldsby Legends* (circa 1840) and in countless pantomimes, cartoons, and live action versions since. It tells of two orphaned children left in the care of their uncle, who passes them onto ruffians to be despatched in order to acquire their inheritance. The two ruffians fall out and kill one another, leaving the children abandoned in a forest. In early versions the children eventually expire due to hunger and cold, their bodies covered with leaves. In some, the children ascend to Heaven, while the wicked uncle receives his comeuppance. The brevity and brutality of the tale mean that it is frequently conflated with a simplified version of Robin Hood to provide a more uplifting narrative arc. Yet, characters have been wandering in and out of the collective story forest for centuries, crossing over into different tales and traditions—as Tolkien acknowledges in his famous essay 'On Fairy-Stories', where he discusses the philological and narratological impulses that drove his legendarium: 'I feel strongly, the fascination of the desire to unravel the intricately knotted and ramified history of the branches on the Tree of Tales' (1997: 120). The tangled roots of story and the assimilatory nature of the oral tradition mean this cross-cultural influence is inevitable, but in much postmodern Fantasy this stitching together of different motifs is deliberate, and often intentionally syncretist or iconoclastic, for example, *Mythago Wood* (1984), *Into the Woods* (Sondheim, 1986), *The Princess Bride* (Goldman, 1999), *Shrek* (Steig, 1990), *The Wild Wood* (de Lint, 1995), *The 10th Kingdom* (Moore, 2000), *Grimm* (2011–2017), *Uprooted* (Novik, 2015), *Silver in the Wood* (Tesh, 2019), and *Gods of the Wyrdwood* (Barker, 2023). Yet this lively cross-fertilisation has been present in taproot texts by Wolfram von Eschenbach, William Shakespeare, Thomas Malory, Jean de la Fontaine, Charles Perrault, the Brothers Grimm, Hans Christian Anderson, William Morris, and many others. Like the far-ranging mycelial networks of the forest, the Fairy Tale is a form that lends itself to synthesis—or a 'parasitic mutualism' (Sheldrake, 2020)—and has thus absorbed and mutated over the centuries, successfully adapting and evolving through the horizontal gene transfer of tropes and forms. As platforms and audiences change, so does the Fairy Tale (Warner, 1995). And the Fairy Tale feeds directly into Children's Fantasy—either directly, through 'straight' retellings as exemplified by Andrew Lang's twenty-five colour-coded fairy books (1889–1913); or by the reconfiguration of motifs and tropes as seen in *Lud-in-the-Mist* (1926), *The Hobbit* (1937), and *The Lord of the Rings* (1954), *The Weirdstone of Brisingamen* (1960), *The Moon of Gomrath* (1963), *The Wolves of Willoughby Chase* (1962), *Where the Wild Things Are* (1963), *The Last Unicorn* (1968), *The Neverending Story* (1979), *and Stardust* (1997), to name a few. One of the most resonant uses of the wood in young Fantasy is in Le Guin's *Tales of Earthsea* (collected in one volume, *The Book of Earthsea,* 2018). The Immanent Grove is situated on the wizards' isle of Roke, and '[i]t is said that the trees of the grove themselves are wise.' This echoes the Ents in *The Lord of the Rings,* but unlike those magical beings there is a sense in Le Guin's Immanent Grove that just by spending time amongst trees one can receive wisdom. Indeed, Le Guin suggests a kind of symbiotic interface in which the human and more-than-human merge: '"How far does the forest go?" Medra asked, and Ember said, "As far as the mind goes"' (2018: 601). This aligns deeply with the Ecological Imaginary, and is made tangible by the Japanese concept of Shinrin-yoku, 'forest bathing', which encourages spending time amongst trees for one's wellbeing (the photochemicals released by the trees are thought to have a beneficial physiological effect, alongside the relaxing effects of light, shade, verdure, and birdsong). Tolkien, as a boy playing with his younger brother, Hilary, around Sarehole Mill on the edge of Birmingham, certainly experienced the benefits of 'forest bathing' without knowing its name. The trees he played on there had a formative influence on him. When his favourite willow tree was cut down, he was devastated—and this early trauma clearly fed into his depiction of the destruction of the woods of Isengard, and the 'last march of the Ents'. The nearby Moseley Bog—now a nature reserve—is thought to have been the inspiration for the Old Wood: the stomping ground of Tom Bombadil, whose Fairy Tale energy seems almost incongruous within the higher register of the Epic of the rest of *The Lord of the Rings* (more in keeping with Tolkien's Fairy Stories, *Roverrandom, Smith of Wootton Major, Farmer Giles of Ham,* and *Letters from Father Christmas*). But it shows how the Fairy Tale 'roots' of story run deep, and cannot easily be removed—they entangle every Children's Fantasy in woodlands, real, imaginary, or metaphorical.

The Wheel of the Year

To a child experiencing the world for the first time, the change in seasons—especially dramatic (remembered) firsts such as the 'first snow', 'first flood', 'first heatwave', 'first storm', etc—can seem momentous, even magical. So it is unsurprising that the seasons, especially seasonal customs and celebrations, loom large in Children's Literature (and by extension, Children's Fantasy). A child's sense of time is different to an adult's. The days and weeks seem longer, nights bottomless, lessons and adult conversations drag, playtime is never long enough, but summers are endless.

Finnish author Tove Jansson (1914–2001) is best remembered for her creation, The Moomins, a hippopotamus-headed family of trolls featuring Moominpapa, Moominmama, Moomintroll, Snorkmaiden, alongside a host of other characters, including the wise philosopher, Snufkin, and the querulous Little My. Dwelling in what appears to be a Secondary World (although it is eventually revealed to be a Wainscot universe connected to this world), Moominvalley, the Moomins experience ecological catastrophe (*The Moomins and the Great Flood*, 1945), apocalypse (*Comet in Moominland*, 1946), winter and isolation (*Moominland Winter*, 1957), and the heights and lows of existence. Through a series of successful, syndicated comic-strips and illustrated books (1945–1970), Tove Jansson not only builds a rich storyworld with its own ecosystem of characters, critters, and locations but also explores material unusual in Children's Literature—melancholy, despair, loneliness—giving the non-anthropomorphic characters psychological realism. The highs and lows of Moomin existence are set against a tangibly seasonal environment—one given substance by Jansson's deep connection to the natural world. As a child, Jansson grew up playing on a small island purchased by her parents; then as an adult she acquired her own small island, Klovharun, with her brother. Here, in the Gulf of Finland, she spent eighteen seasons (April–October) with her lover and companion, Tuulikki Pietilä—writing, drawing, filming, sculpting, swimming, fishing, and playing games of dressing up and make-believe. These experiences fed into her Moominverse, but also manifested in her adult fiction—most notably *The Summer Book* (2020 [1972]). This is not Children's Fantasy in the strictest sense, but a sense of the power and importance of fantasy permeates it. It relates a bittersweet tale of an endlessly curious and questioning six-year-old girl, and her artist grandmother, spending a summer on a small island off the coast of Finland. A wry quality inflects the prose—chapters brimming with a restless, ludic energy. The little girl continually makes up games with her grandmother, who unbeknownst to her, is dying. The realism of the setting and the convincing dynamic between grandmother and granddaughter is not surprising—for *The Summer Book* drew upon childhood memories of her beloved family island, and also memories of her mother and niece, Sophia Jansson. The book was written shortly after the death of the author's mother—and is an indirect tribute to her. Tove's grief found expression here, and perhaps some kind of healing or, at least, acceptance: the death of the grandmother synchronises with the end of summer. The child must learn that all things end, especially the things we love most. It is this hard edge of existential truth interwoven with the charming exploits of the Moomins and their neighbours that has won over so many adult readers to Jansson's oeuvre. Through her books, millions of young readers have learned about nature, and their own nature—the seasons act as an extended pathetic fallacy, but also a reminder of the inevitability of change, growth, and decay. Susannah Clapp suggests Jansson's work is increasingly topical:

> 'For all the light clarity of the prose—which is comic, benign and quizzical—these books show places gripped by ferocious forces, laid waste by storms and floods and snows. They speak (but never obviously) of characters resonating to the winds and seas around them. They include visions that now read like warnings of climate change.'
>
> (2021)

Jansson's books provide a resonant attunement to nature, and psychological preparation for its extremes.

Yet sometimes the seasons bring transformation and enchantment.

The Box of Delights, or *When the Wolves Were Running* (1935) is a Children's Fantasy novel by John Masefield (UK Poet Laureate 1930–1967). It is a sequel to The Midnight Folk (1927) which first introduces the young protagonist, Kay Harker, and features a coven, talking animals, and the formidable governess, Sylvia Daisy Pouncer, and the wizard Abner Brown—who reappear in the sequel, up to more nefarious deeds. *The Box of Delights* is set over a particularly snowy Christmas, with the English countryside transformed into a landscape of magical immanence. It was made into a very successful and much-loved BBC TV series, broadcast over Christmas 1984, featuring Patrick Troughton (the 2nd Doctor Who) as a centuries-old cunning man, Cole Hawlings, Patricia Quinn as Sylvia Daisy Pouncer, Robert Stephens as Abner Brown, and Devin Stanfield as Kay Harker. The production, directed by Renny Rye and adapted by Alan Seymour, won several BAFTAs and RTS awards and featured special FX that were cutting-edge for their time (and, costing one million pounds, made it the most expensive children's series produced on British television to date).

The opening credits are deeply eerie—behind-the-sofa stuff, enhanced by the haunting, magical theme tune combining 'The First Nowell' with the third movement of the 'Carol Symphony'.

Masefield's book influenced the burgeoning imaginations of many a young reader, and it paved the way for later Children's Fantasy informed by the darker tones and social realism that started to inflect such works in the post-war era, especially from the 1970s, when a sharper edge started to appear.

The Dark Is Rising by Susan Cooper was published in 1973 and is the second in a sequence that became known as the same. It is set in the Thames Valley and tilts it, via heavy snowfall, into an estranged, enchanted land, as *The Box of Delights* does with the Malvern Hills in Herefordshire. Like Masefield's novel, it too features Herne the Hunter, and a young protagonist getting inveigled in a Manichaean fight between the forces of the Light and the Dark. Instead of a magical box, the young Will Stanton becomes The Sign Bearer, acquiring six ancient signs that help him in his battle against the Dark that assails his village and community (in *His Dark Materials* Lyra Belacqua becomes the 'Sign Reader' with the alethiometer—a device that enables the user to discover the truth). Will discovers the presence of The Old Ones in his neighbourhood—The Blacksmith, John 'Wayland' Smith; The Rider; The Hunter; The Lady, Mrs Thorncrofte; Merriman Lyon; and The Walker, aka Hawkin (who is like a shadow version of Cole Hawlings), among others—before it is revealed that he is the last of them, with a key role to play in the final battle. The River Thames is a dominating presence, and the land around it floods, as happens in *La Belle Sauvage* (the first in Pullman's long-awaited follow-up to *His Dark Materials, The Book of Dust*). Cooper presents us with a vision of a flooded, rewilded England, echoing Richard Jefferies' 1885 post-apocalyptic novel, *After London: or Wild England*. As with Alan Garner's Alderley Edge, the landscape around Will's village becomes a crosshatch of the magical and the mundane. An old track becomes The Old Track, a fact which saves Will Stanton from The Rider. The magical maelstrom that erupts is navigated by knowledge of deep place, and by the bonds of community.

During the winter of 2021, the writer Robert Macfarlane instigated an online read-along on Twitter. The book had long been a favourite of his—reading it in childhood, he has then gone on to introduce it to his children, and to bring it the attention of a whole new generation of readers and listeners, for in the following year, he worked on an adaptation for the BBC World Service, produced by BBC Sounds/Theatre Complicité, which was broadcast to synchronise with the Christmas setting of the novel (like *The Box of Delights* 1984 TV series). The twelve fifteen-minute episodes chimed well with the festive holiday, being released daily at breakfast time (GMT), cultivating a shared listening experience akin to when *The Hobbit* and *The Lord of the Rings* radio adaptations were broadcast in the 70s, resulting in whole campuses practically grinding to a halt, as everyone listened in. The sound FX created by the BBC Radiophonic Workshop vividly brought to life the wintry, spectral ambience of the book—the supernatural 'thickening' that occurs as the apparent reality is destabilised by increasingly powerful intrusions—along with the resonant theme tune written and performed by Johnny Flynn. *The Dark is Rising* is Intrusion Fantasy at its best (to use Mendelsohn's speculative taxonomy). Its effectiveness is thanks to a combination of a tangible setting, evocative seasonality, and the archetypal and the actual in the interweaving of the mythic, folkloric, and religious with the comforting familiarity of Christmas.

Animals and Other Critters

The more-than-human is a constant fascination to children, for whom animals embody an (apparently) inarticulate alterity which expresses something of their own animal-self—driven by instinctual needs for sustenance, safety, affection, companionship, and competition. To Abram, encounters with animals (as well as with rocks, trees, rivers, grasses, etc.) remind us of what we have forgotten—but perhaps children in their proprioceptive beingness and neo-animist consciousness, have not: 'We've largely forgotten the incantatory and invocational use of speech as a way of bringing ourselves into deeper rapport with the beings around us, or of calling the living land into resonance with us' (2010: 174). He argues in 'The Speech of Things' that *everything* communicates—and not just what we consider sentient and intelligent—through not only sound, but also scent, movement, colour, chemical signals, and countless subtle forms we are only now becoming privy to through the emerging science of biosemiotics: how life speaks to life. Abram calls this the 'articulate landscape . . . a community of expressive presences that are also attentive and listening, to the meanings that move between them' (2010: 173). He argues that indigenous oral peoples share a common sense of this. However, generalising this claim (to say there is one paradigm or umwelt that sums up all tribal societies is homogenising and essentialist), it could be suggested that many children seem too also—although this may be decreasing as access to natural spaces diminishes, for it is an ecoliteracy cultivated in situ. Zoos and urban farms can only go so far. Nevertheless, it is a modality that children readily accept—talking animals, trees, springs, flowers, insects, etc., proliferate in Children's Fantasy.

In *The Box of Delights* (1935), animals play a prominent part—a flying white horse, wolves, stags—with characters turning into them (Herne the Hunter) or existing as unsettling hybrids (e.g. Pirate Rats). In the *Worzel Gummidge* books by Barbara Euphan Todd, published 1936–1963, the titular character is conversant with the natural world. This is leaned into in the most recent TV adaptation, written and directed by Mackenzie Crook (2019–2021), in which the beauty and biodiversity of the English landscape is foregrounded in terms of both imagery and plot—the evocative visuals and gentle folk soundtrack seduces the viewer into the inherent glory (and thus preciousness) of nature. Gummidge and his magical community of fellow scarecrows are portrayed as part of an endangered demi-monde—with their progenitor, The Crowman (played as an eccentric tramp by Michael Palin) is a marginalised figure scraping a living laying hedges. And awareness of Climate Change and threats to biodiversity and habitat play a prominent role in the plot. Seasons are out of synch, and nature is out of kilter. And yet enchantment, for a while, remains—charged with a poignant evanescence.

In *Watership Down* (1972) Richard Adams not only enables the reader to fully inhabit a non-anthropomorphic reality, via a warren of rabbits on the Berkshire Downs, but also introduces a sobering note of environmental sensibility—an 'eco-realism' that had become increasingly prevalent since the publication of *Silent Spring* (Carson, 1962). The novel influenced the nature-writer Nicola Chester, who went on to become a regular contributor to the nature diary of the *Guardian,* and author of her nature memoir, *On Gallows Down* (2021).

And yet, Children's Fantasy does not have to be mimetic (that is set in this world, or appearing to represent the consensus reality) to have influence. Much-loved classics such as *The Horse and His Boy* (Lewis, 1954), *The Last Unicorn* (1968*), and The Neverending Story* (Ende, 1979) draw upon the hippophilia that many children (especially, but not only, girls) experience.

Drawing upon the oral and literary traditions of the animal tale and the parable are books such as Rudyard Kipling's *Just So Stories* (1902), which were initially told as bedtime stories to his daughter, Effie (Josephine). Kipling explains their origin: 'in the evening there were stories meant to put Effie to sleep, and you were not allowed to alter those by one single little word. They had to be told just so; or Effie would wake up and put back the missing sentence. So at last they came to be like charms, all three of them—the whale tale, the camel tale, and the rhinoceros tale' (cited by Karlin, 2015). *The Jungle Book* (1894), with its cast of talking animals, also draws upon the legacy of Aesop, along with Indian myths and legends. This literary tradition has not diminished in popularity—joining best-selling modern classics like *Fantastic Mr Fox* (Dahl, 1970) and *The Gruffalo* (Donaldson and Scheffler, 2017), there is *The Boy, The Mole, The Fox, and The Horse* (2019) by illustrator-author, Charlie Mackesy—although in its dispensing of life lessons the latter follows more in the flight-path of former aviator Richard Bach's vastly popular *Jonathan Livingstone Seagull* (1972).

However, Bach's novella has a more transcendentalist tone, perhaps influenced by the Eastern Mysticism popular in the late Sixties. Nevertheless, John Clute includes it in *The Encyclopedia of Science Fiction,* describing it as 'an animal fantasy about a philosophical gull who is profoundly affected by flying, but who demands too much of his community and is cast out by it. He becomes an extremely well-behaved accursed wanderer, then dies, and in posthumous fantasy sequences—though he is too wise really to question the fact of death, and too calmly confident to have doubts about his continuing upward mobility—he learns greater wisdom. Back on Earth, he continues to preach and heal and finally returns to heaven, where he belongs' (Clute, John, and Peter Nichols, 1993: 79).

There are many other examples of such sites, and we shall look at a good cross section in the 'Fantasy Sites' chapter. For now, we must leave the mythopoeic landscapes of childhood and step into the worlds of Epic Fantasy.

3 Epic Fantasy

Portals, Paradises, and Wastelands

Epic Fantasy, by its very nature, foregrounds the setting—typically, a Secondary World—which provides a backdrop to a series of interconnected narratives, sometimes taking place simultaneously in dramatically different terrain. One only needs to think of *The Lord of the Rings*, or *A Song of Ice and Fire*, to get a quick sense of what an Epic Fantasy feels like (at least in its mainstream form). It may be the most familiar kind of Fantasy, as it has proved to be commercially successful with respective 'franchises' attracting a legion of fans. John Clute defines Epic Fantasy as '[a]ny fantasy tale written to a large scale which deals with the founding or definitive and lasting defence of a LAND may fairly be called an EF' (1999: 319). Yet he problematises it when he adds the caveat, 'Unfortunately, the term has been increasingly used by publishers to describe HEROIC FANTASIES that extend over several volumes, and has lost its usefulness' (1999). Clute was perhaps thinking of Robert Jordan's *The Wheel of Time* (1990–2013), and the whole slew of similar titles that have jumped on the post-Tolkienian bandwagon. Allen Stroud has since attempted to rehabilitate what Le Guin somewhat disparagingly categorises as 'Commodified Fantasy' (*Tales of Earthsea* foreword, 2018 [2001]), arguing that it has its place in the cultural ecosystem, and can innovate from within: 'it is by no means the case that all Commodified Fantasy is badly written, and its stereotyping performs a useful function in providing the genre with an anchorage and a steady sales base' (Stroud, 2023: 107).

In his definition of Epic Fantasy (2023: 152–153), Stroud also acknowledges the Classical origins of Epic Fantasy: the long narrative poem that draws upon mythological material, usually focusing on the exploits of legendary heroes—a form best exemplified by the works of Homer (*The Iliad* and *The Odyssey*), Apollonius of Rhodes (*The Voyage of Argo*), and the anonymous *Epic of Gilgamesh*, as well as works of collective authorship such as the Hindu epics, the *Mahabharata* and the *Ramayana*, and the Tibetan and Central Asian cycle, *The Gesar of Ling*. In his *Poetics* (1907), Aristotle was the first to formulate a definition of the Epic, noting that

DOI: 10.4324/9781003463665-4

it was a verse form that focused on 'characters of a higher type', like its cousin, Tragedy, but seeing it as inferior in form (although 'all the elements of an Epic poem are found in Tragedy'), with the singular characteristic of not being bound to a 'single revolution of the sun', and can thus cover a vast span of history: 'the Epic action has no limits of time'. So, even in Aristotle's culturally specific definition, we see the elements of the Epic Fantasy familiar to us—the focus on dramatic narrative, the long sweep of history, the characters of a 'higher type'. What emerged as an oral form became, even with Homer, a literary form—albeit one that evokes the orality of its origins (Lord, 1976). By the 16th century, poets such as Edmund Spenser and Ludovico Ariosto were self-consciously adopting it to create a higher register suitable for a contemplation of transcendental themes—a highly intertextual and sophisticated form that required a knowing audience, and thus very flattering to the courtly audiences it was created for. Stroud goes on to acknowledge how Tolkien's very deliberate evocation of the form led ultimately to its derivation and commodification:

> 'J.R.R. Tolkien's lifelong dalliance in Middle-Earth was conceived as an attempt to synthesize the epic that Old (i.e., pre-Norman Conquest) England never had; insofar as *The Lord of the Rings* is spun off from *The Silmarillion*, the primary model of commodified fantasy retains and exemplifies many of the pretensions as well as the narrative formula of the epic. This makes the notion of "epic fantasy" more than a mere advertising slogan, although its frequent use as such has reduced its critical utility somewhat.'
>
> (2023: 153)

Yet, as a catch-all term for a substantial amount of sometimes highly significant and influential Fantasy texts, which between them chart a vast amount of Secondary World imaginary 'landscape' (from Secondary Worlds to whole cosmologies), Epic Fantasy, has validity for our current purposes, especially in its connection with the concept of 'The Land'. Clute describes this as 'a land may be defined as a secondary-world venue whose nature and fate are central to the plot: a land is not a protagonist, but has an analogous role. Some or all of the following will almost certainly be the case: the land may evince WRONGNESS; it may be subject to THINNING; it may be a WASTELAND; it may suffer (or be saved by virtue of) a fundamental TRANSFORMATION; it can be healed; and it is almost certainly, in some sense, alive' (1999: 558). Rather than dwell on the specificity of each of the terms Clute refers to here, they shall be touched upon within a wider discourse around the Weberian axis of enchantment–disenchantment–re-enchantment, which will be the main focus of this chapter.

Before we get into the granularity of that, it has to be acknowledged that the very idea of 'The Land' in Fantasy has become something of a cliché, but that is arguably because the notion of a landscape of heightened significance—metaphoric; ensouled; reflective of the psychic health and moral atmosphere of its inhabitants—is intrinsic to the genre. It is a continuation of the mythscapes of childhood, the 'cottages of lost play', which Fantasy fans like to return to, at least for a while; but also, perhaps unconsciously, a desire for a neo-animist connection with the natural world, which much of the technological and remediated texture of modern life feels antithetical too. The enchanted landscapes of Fantasy offer a correlative to this—a soothing counterspell; or, increasingly, an echo and ironic counterpoint in the disenchanted, compromised vistas that have gained popularity in late 20th-century/early 21st-century iterations where mundanity, geopolitics, and the environmental crisis have seeped in.

The landscapes of Fantasy were no longer the reliable refugia of old.

From the mid-20th-century onwards in the wastelands depicted in *The Dying Earth* (1950–1984), *Elidor* (1965), *The Chronicles of Thomas Covenant the Unbeliever* (1977–1979), *Viriconium* (1971–1984), *The Neverending Story* (1979) and others suggest a kind of existential crisis over the nature of Fantasy and our relationship to it. Our dreams were no longer safe, but had been corrupted by the counter-discourse of reality. This transference was perhaps inevitable, and arguably always there: Tolkien's Middle-Earth could not have existed without the Industrial Revolution and the Somme. And further back, the felling of the sacred cedar wood in The *Epic of Gilgamesh* epitomises the development of civilisation at the expense of nature. The real provides the coal for the fires of our imaginary worlds, even as it pollutes the air we shall breathe there.

Before we consider its often compromised quality in more detail, let us return to its formation and most influential iterations.

The Land

Firstly, let us consider the idea of 'The Land' in more detail. This has become such an overused trope within 'Commodified Fantasy', although Sangster critiques this and, in his robust defence, 'The Value of Iteration' (2023: 97–161), is at pains to rehabilitate it as a place for a lively engagement with, and often inventive reworking of, many of the tropes of the genre. However, any trope can atrophy into the formulaic and the over-familiar, and we stop noticing it, for example, the enchanted forest of Secondary World fantasy; the grim tech-noir metropolis of cyberpunk, etc. Although more recent manifestations of 'The Land' have started to weave in implicit or explicit environmental elements—such as the drought and famine in *The*

Summer Tree (2006 [1985]), the toxic wasteland of *Viriconium* (2000), and the eco-catastrophe of *The Silence in The Fifth Season* (2006)—early Fantasy often sought to depict a space of ontological palimpsest, where this world and 'otherworlds' overlapped. The otherworld was glimpsed at 'thin places', where the porousness of the worlds was most pronounced—ancient, sacred places; small islands; springs, pools, and lakes; forest groves; remote tracts of wilderness; and threshold times, such as dawn, dusk, midnight, full moon, May Eve, Midsummer, Hallowe'en, and New Year's Day. Protagonists often transitioned from one to the other without realising it, or encountered supernatural beings unexpectedly, as in the first of the *Four Branches of The Mabinogion/Y Mabinogi* (Guest, 2000) where Pwyll, Lord of Dyfed, who when out hunting with his hounds runs to ground a splendid stag, only for his quarry to be claimed by another hunter, Arawn, Lord of Annwn (the Celtic Underworld), who asserts that it was *his* pack–identified as fairy hounds by the white coats and red ears—which had cornered it. Another time, Pwyll stands upon the mound of Arbeth (Narbeth, Pembrokeshire), where it is said one may either receive a wound or behold a wonder. From it, Pwyll beholds a beautiful lady ride by on a fine horse, accompanied by a magical host of birds. He sends a servant to discover her name, but no matter how fast the retainer rides, he is unable to catch up with her, although she seems to just canter. When he descries her again, Pwyll himself gives chase, but encounters the same problem, until he calls out, 'Lady, in the name of the one you love best, I pray thee stop!' This does the trick, for it is Pwyll himself, Rhiannon (as she is called) her heart as chosen.

In the west of Ireland, the poet Oisín and his father were out hunting with their favourite hounds, when they encountered a similarly resplendent and irresistible fairy woman, Niamh of the Golden Hair, who had come from Tir nan Og, the Land of the Ever Young, to wife the poet, whose golden words and music had carried to her enchanted shores. She persuaded him to return with her, although this was at great cost—for her island was a 'chrono-abyss' (Clute and Grant, 1999: 946–947), and when Oisín returned to Erin only a few weeks later, he discovered three hundred years had passed and all he had known and loved had turned to dust. In another Hibernian myth, 'The Children of Lir', it is the young protagonists themselves—the four children of the titular king—who are translated into a magical ontology when they are turned to swans by their stepmother, Aoife. They are cursed to remain in that form until the 'Man of the North weds the Woman of the South', and to roam the five quarters (the fifth being the sacred centre of Tara) of Erin, spending three hundred years on Lake Derryvaragh, three hundred on the Straits of Moyle, and three hundred by the Cliffs of Moher.

Such supernaturalised landscapes act as places of testing and attentiveness, karmically accelerated places where every word, gesture, and deed has significance, and everything encountered is deeply metaphoric, although sometimes in an inchoate, oneiric way. In the tale of Peredur in *The Mabinogion/Y Mabinogi*, the questing knight at one point comes to a vale through which meanders a shallow river. On one side, there is a flock of white sheep, on the other, black. Whenever a black sheep bleats, a white sheep crosses the water and turns black, and vice versa. In 'The Countess of the Fountain', from the same source, the errant knight, Yvain, hears of a sacred spring by a great tree. By the spring there is a rock, and upon the rock there is a silver dish attached by a chain. Whoever seeks adventure (as errant knights are wont to do) has to fill the dish from the spring and dash the water upon the rock, and then storm clouds gather, thunder rolls in the heavens, and a black knight appears, who must be met in combat. Robin Hood encounters a similar moment of reckoning, when crossing a narrow bridge over a merry brook his way is blocked by Little John, wielding a quarterstaff. Before they became friends, they fought as foes. When Robin Hood comes off the worst, ending up in the water, Little John is amused and placated.

Forests in Fantasy are frequently sites of trial, tribulation, and trickery. The broken sightlines and difficulty in orienteering they create; the concealment and sudden revelation they facilitate. They are a gift to storytellers, as Shakespeare and many others have realised. In the Arthuriad, the 'Forest Perilous' is a heightened zone of adventure where knights can test their mettle. In an early story about the life of Merlin, the magician is driven 'mad' by the witness of a massacre and retreats to the forest of Caledon, in the Scottish Borders, where he becomes a nature hermit known as Myrddin Wyllt ('Wild Merlin'), communes with animals, and has an extended conversation with a pig—which sounds eccentric, but Merlin would have known pigs were believed to hail from the underworld and were therefore wise. In the Cornish legend of Tristan and Isolde, the two lovers flee to a wild wood to escape the wrath of King March (who Isolde had been betrothed to). King Arthur's nephew, Gwalchmai, the 'Hawk of May', is despatched to find the erstwhile knight and bring him home. Through a long colloquy, the knight is rehabilitated into human society, he is 're-membered', in the way Yvain is after he goes feral through shame at the way he has treated a kind handmaiden who helped him woo the Countess of the Fountain. Yvain lives in the wilderness, half-man, half-beast. He is befriended by a lion, which helps him time and time again. Finally, the knight is restored to health and sanity by the love of the handmaiden. In all of these stories it seems that any kind of imbalance can be redressed in nature, and by a return to a natural state of untainted vision and heart.

Parsifal, who is raised in ignorance of his noble birth in a forest, eventually wins the Holy Grail through the purity of his nature and cleansed perception—he literally 'perceives the veil', a destiny encoded in his name, and asks the Grail Question (Matthews) that heals the perpetually ailing Fisher King. It is suggested that the King and the Land are one, and when the former becomes diseased or wounded the latter becomes wasted. In the story of another Fisher King figure, Amangons, thirsty travellers were once slaked by the 'Maidens of the Wells', who offered them a drink from a golden chalice that was wonderfully refreshing. Hearing about this, Amangons, the king orders his knights to seize the maidens and the magical cup. He wants to hoard the priceless power of the cup for himself, but its gift taints in his possession—he sickens, but is unable to recover, or die. His extractivist policy has backfired.

In many folk tales, myths, legends, and fairy tales a similar ecological morality exists—anyone who asks selfishly and seeks to hoard a sacred treasure for themselves pays the price. True magic, when it happens, is given freely, and never bought, sold, or exploited. It is a kind of grace bestowed beyond the quid pro quo reciprocity and rapaciousness of men. In numerous stories, when the Land is mistreated, its denizens and spirits rise up. We are warned to treat it well.

The Tour

It is no coincidence that some of the earliest literary Fantasies emerged during the same period (15th-17th centuries) as the so-called Golden Age of Exploration (it was less gilded for the indigenous peoples who were invaded, massacred, infested, enslaved, and subjugated). As stories of the New World and other 'exotic' locations fired the popular imagination—increasingly wild tales of unimaginable bounty and beauty, attractive or fearsome natives, strange and terrible creatures, and untold riches—the literary arts started to reflect this. Thomas More's unattainable island paradise, *Utopia* (1516); Edmund Spenser's flight of fancy, the epic Elizabethan poem, *The Faerie Queene* (1590–1596); and Ludovico Ariosto's *Orlando Furioso* (1506–1532), to name a few—the latter, ostensibly a chivalric romance set in the time of Charlemagne, yet featuring a trip to the moon, a giant sea monster called Orc, a hippogriff, and the unrequited love between the paladin Orlando and the Pagan princess, Angelica. Perhaps the quintessential 'New World Fantasy' amongst these is William Shakespeare's final completed play, *The Tempest* (1610–1611). Survivors of a shipwreck find themselves on a remote island ruled over by a magician, Prospero, who is attended by an otherworldly creature called Caliban. His beautiful daughter, Miranda, reared in ignorance of the ways of men and the world, becomes a much sought-after object of desire; but her father's godgames ensure there is no smooth path to fulfilment here. Famously relocated to deep space as the classic science fiction film, *Forbidden Planet* (dir. Wilcox, 1956) and adapted countless times in diverse media, often in very inventive ways (e.g. Peter Greenaway's 1991 *Prospero's Books;* or Julie Taymor's 2010 version starring Helen Mirren), *The Tempest* has provided rich soil for scholars wishing to explore orientalism and postcolonialism, with the maligned, marginalised 'native' Caliban at the heart of much of the discourse. Even in narratives ostensibly closer to home, such as *A Midsummer Night's Dream,* set in a 'wood near Athens', the exoticisation of the other serves as a way of connecting the familiar with the fantastic. As with the countless travellers' tales spawned by the Golden Age of Exploration, the frisson of possibility combined with the gullibility of the audience (and voluntary credulity—because everyone loves a good tale), fermented a potent mix of ontological potentiality: the very stuff that fantasy is made on. So, the Swan of Avon was not alone in conflating 'foreign' with Faerie—in many ballads and folk tales, the otherworld is referred to simply as a faraway country: any name will do, from Ultima Thule to Van Diemen's Land. Often this mythogeographical fabulation—Hy Brasil and Eldorado, to name two examples—was used as apologia for the brutally exploitative programme of conquering and colonisation. Conquerors and colonists set out in search of these places, and so they found them, projecting their rapacious desire (and 'manifest destiny' entitlement) onto them. They were the Chosen Ones, and this was the Promised Land. This colonialist gaze can often be detected in certainly the early examples of Fantasy, although sometimes with a sardonic or ironical distance, as in William Beckford's *Vathek* (1786), Lord Dunsany's *The Gods of Pegāna* (1905), Robert W. Chamber's *The King in Yellow* (1895), E.R. Eddison's *The Worm Ourobouros* (1922), Hope Mirrlees' *Lud-in-the-Mist* (1926), and Sylvia Townsend Warner's *Kingdom of Elfin* (1977).

This colonising tendency reappears in later, Tolkienesque commercial fantasy, but often without the self-reflexivity. The Secondary Worlds of Epic Fantasy became the setting for what Diana Wynne Jones mockingly refers to as 'The Tour' (2004: 213–214). Provided with a map at the start of the book, we can safely assume we'll be visiting most of the places featured on it. The viewpoint is often that of the outsider—sometimes from our world, whose sensibilities are perhaps a bit closer to our own, thus providing a relatable counterpoint to the 'exotic' locals, culture, and customs encountered. Such works are the cousins of the accounts of the 'Grand Tour' popular with the privileged elites of the 18th and 19th centuries. Even worse, the (usually) white, male, Western protagonist is destined to become the 'redeemer' of this land and its people. The destinarianism of these narratives feeds into a Messiah Complex, which is (partly) redeemed by the trope of the 'reluctant messiah', although any such prevarication or outright denial on the part of the 'chosen one' is usually seen as a good sign, and that they are *really* The One. Stephen Donaldson's *The Chronicles of Thomas Covenant the Unbeliever* (*Lord Foul's Bane*, 1977; *The Illearth War*, 1978; *The Power That Preserves*, 1979, collected 1993) were effective at deconstructing this expectation, with a leprous anti-hero. Later works of Fantasy have mindfully interrogated this trope, but perhaps thanks to the popularity of roleplaying games and extended adventures known as 'campaigns' (which are sometimes remediated as Fantasy novels), the idea of The Tour lingers in less innovative Fantasy novels. If there is a map, we know we are going to get a lot of geography, a detailed itinerary, and an awful lot of overly demonstrative world-building, as Wynne Jones wittily points out: 'Find the Map. It will be there. No Tour of Fantasyland is complete without one' (2004: 1). Such attentiveness to the setting *can* ensoul and ennoble a landscape, and may make us even look at our own, real-world environments, with enriched perception, but at its worst, it can be like looking at someone's holiday snaps.

However we choose to depict the worlds of our Fantasy, every iteration is a possible confirmation of the Sapir-Whorf Hypothesis—the language we use to speak about the world influences how we perceive it. If a landscape is 'merely' a backdrop, with only the minimal amount of texture sketched in, an annoyance or inconvenience to be crossed as quickly as possible, then it encourages readers to perceive it in that way too. Perhaps Tolkien, with his love of walking, of cartography, and of the natural world, went too far in the other direction, then it is surely preferred to its opposite: where many modern young Fantasy fans prefer to 'skip' the description in their reading and in their writing—resulting in thin prose, an impoverished literary experience, and denuded imaginations. In terms of ecolinguistics, the language we choose (or do not choose) to talk about the world with fundamentally affects the way we value it.

But now let us turn our attention to the main critical vortex at the heart of Fantasy landscapes, where this can be seen to be very much in action.

Paradise Found

Max Weber argued that, since the Enlightenment and the Age of Reason—which saw the development of rationalism and the scientific method, an empirical, materialistic world view that sees the universe as mechanistic (as opposed to the religious and spiritual)—the world has gone through a process of disenchantment: *Entzauberung*, literally 'de-magic-ation'. This has left the world a poorer place in Weber's mind. Like many recent Fantasy worlds, our world experienced a 'thinning': aesthetic, sociological, ontological. Weber's theory has been critiqued by those who still find mystery and wonder in the world, indeed in science itself and the infinite complexity of the universe. But one can see how, in reaction to this general trend of disenchantment, a counter-narrative emerged—one that was not necessarily anti-Enlightenment, just pro-Enchantment: a recalibration of reality, one that was essential for psychic health, for it acknowledged the dreams, nightmares, visions, and fantasies that are as much a part of the human condition as our base physiological needs and highest, noblest thoughts. Consciousness, and our innate beingness in the world, is not purely biological and locked into the realm of the senses. Human beings are more complex than that. As are the other-than-human and all that we share this biosphere with. The more we discover, the more this is confirmed. Scientists working at the cutting-edge of their field (such as mycologist Merlin Sheldrake, bryologist Robin Wall Kimmerer, and biologist David George Haskell) frequently express their profound awe and humility in the face of what they unearth. And yet, confronted with centuries of narrative to the contrary, it is small wonder that a growing body of literature emerged that offered an antidote to what T.S. Eliot phrased as 'too much reality'. According to the High Priest of Modernism, 'humankind cannot bear [it]' (*The Four Quartets*, 1943). Even as he charted its squalid texture with his cool gaze, Eliot acknowledged the ghost in the machine, 'the heavy burden of the growing soul' and the epiphanic promise of 'shafts of sunlight', which for him ultimately led back to God through the Anglican faith. And so, the collective works of Fantasy written over the centuries since the Enlightenment could be seen as a collective attempt to hold back the tide of 'too much reality', a sea wall that continually sprung leaks, and was perhaps inevitably going to burst at some point, for these polders of the imagination always live on borrowed time. We shall look at that in more depth in the next section.

For now, let us evaluate the endeavour of enchantment, which is discussed in an early essay by Joseph Addison—one of the very first attempts (David Sandner calls him the 'first critic of the Fantastic', 2011) to justify the project of Fantasy at a time when rationalism was on the ascendent, and any kind of supernaturalism and overt religiosity was dismissed as belonging to the past. Joseph Addison was to provide a 'working definition' of the Fantastic for the long 18th century in his 'fairy way of writing'—forged from the contemporary tensions at the fault-line between religious belief and positivist scientific discourse:

> Thus we see how many ways Poetry addresses it self to the Imagination, as it has not only the whole Circle of Nature for its Province, but makes new Worlds of its own, shews us Persons that are not to be found in Being, and represents even the Faculties of the Soul, with her several Virtues and Vices, in a sensible Shape and Character.
>
> (1712)

Addison goes beyond the 'inherent values' of 18th-century personification in his conception of the Fairy way of writing as not restricted to 'mere allegory' (and in this sense his argument is proto-Tolkienian), and imagines the function of such writing in making 'new Worlds of its own' and other ways of Being. Commenting on this influential essay, Sandner goes on to emphasise the dialectical role of the Fantastic—suggesting it has always been in dialogue with the Weberian *Entzauberung*: 'The fantastic is an acutely modern literature always coming after primary belief in the supernatural has been emptied of meaning, marking the uncanny return of exploded supernatural belief' (Sandner, 2011: 12). You cannot have enchantment, it would seem, without the lack of it, which makes each attempt to evoke it a poignant narrative of absence and loss, pre-empting the solastalgia of the Anthropocene.

The Medieval imaginary land of Cockayne is a land of plenty born of extreme hardship—poverty, continual hunger, disease, war, famine, and low life expectancy, often brought about by back-breaking work in a punitive feudal system designed to subjugate and exploit the many in favour of the few. In Cockayne, rivers flow with wine, fowl and fish offer themselves up freely, loaves grow like leaves, fruit is always in season, and the status quo is topsy-turvy: abbots are beaten by their monks, nuns saucily offer up their bottoms, the serfs snooze, and the worst off are free of want. Such a scene is depicted in Pieter Bruegel the Elder's vision of The Land of Cockayne, *Luilekkerland* (1567), where well-fed peasants sleep off their feasting and carousing amid their discarded tools, a table tilts under the weight of fine food, houses are made of cake, and pigs run around with a knife obligingly strapped to them, ready to be butchered. This is Fantasy of the common folk, and has unsurprisingly survived into the present day. In the depths of the Dust Bowl of Depression-era America, the Land of Cockayne became the 'Big Rock Candy Mountain' in the 1928 song by Harry McClintock, a hobo's dream where cigarettes grew on trees and hens laid soft-boiled eggs. In his *Jurgen: a Comedy of Justice* (1919), American writer James Branch Cabell has the land of Cocaigne between the lands of sunrise and morning: 'Thus Jurgen abode for a little over two months in Cocaigne, and complied with the customs of that country. Nothing altered in Cocaigne: but in the world wherein Jurgen was reared, he knew, it would by this time be September, with the leaves flaring gloriously, and the birds flocking southward, and the hearts of Jurgen's fellows turning to not unpleasant regrets. But in Cocaigne there was no regret and no variability, but only an interminable flow of curious pleasures, illumined by the wandering star of Venus Mechanitis' (Chapter 24). Roald Dahl's popular Children's Fantasy *Charlie and the Chocolate Factory* (1964) is only one of many post-war fantasies of plenty, fuelled by the hardships of rationing and infantile dreams of abundance.

Yet beyond these fantasies of sugar-craving and base desire, narratives of enchantment often articulated more existential or transcendental needs. Enchanted islands such as Tir nan Og (The Land of the Ever Young) or Avalon (where King Arthur, the 'once and future king', goes to heal of his grievous wounds) offer us consoling fictions about escaping death, remaining ever youthful, or healing of our hurts. In her sales pitch to the poet Oisín about her magical island, Niamh of the Golden Hair emphasises these qualities:

"Delightful is the land beyond all dreams,
Fairer than aught thine eyes have ever seen.
There all the year the fruit is on the tree,
And all the year the bloom is on the flower.
"There with wild honey drip the forest trees;
The stores of wine and mead shall never fail.
Nor pain nor sickness knows the dweller there,
Death and decay come near him never more.
"The feast shall cloy not, nor the chase shall tire,
Nor music cease for ever through the hall;
The gold and jewels of the Land of Youth
Outshine all splendours ever dreamed by man.
"Thou shalt have horses of the fairy breed,
Thou shalt have hounds that can outrun the wind;
A hundred chiefs shall follow thee in war,
A hundred maidens sing thee to thy sleep.
"A crown of sovranty thy brow shall wear,
And by thy side a magic blade shall hang.
Thou shalt be lord of all the Land of Youth,
And lord of Niam of the Head of Gold."

Not only does Niamh's poem outline its many attractions beautifully, but it is also a performative speech-act in its own right, creating the effect of enchantment in its declamation and seducing Oisín into accompanying her. As a poem performed orally during a rendition of the Irish myth, it can have a similarly spellbinding effect. Tir nan Og is, in effect, the ultimate luxury resort. And yet a stay there will cost you dearly—all that you have known and loved will have died and decayed by the time you return. You will not recognise your homeland, becoming a superannuated anachronism as soon as you set foot on the earth (or touch a stone in Oisín's case). As with Odysseus on Ogygia or Circe's island, the crew of Bran or St Brendan getting entangled on the magical islands they encountered on their echtrai and immrama, or Captain Kirk getting over-familiar with an attractive alien maiden, there is always a cost. A dark secret or a fatal flaw in one of its inhabitants is revealed—as in the stories of doomed paradises such as Ker-Ys, Lyonesse, Numenor, and Huxley's *Island.*

In H. Rider Haggard's 'a history of adventure', *She* (1887), a formidable, beautiful queen, Ayesha, known as 'She-Who-Must-Be-Obeyed', rules over a lost African kingdom, preserved in her immortal beauty by the Pillar of Fire, until her love for Leo, whom she believes to be her re-incarnated lover, makes her pay the price. Sometimes it is impossible to stay in these enchanted polders for purely practical reasons, as in James Hilton's *Shangri-La* (from his 1933 novel, *Lost Horizon,* adapted from Frank Capra in 1937), a remarkably comfortable and cultured lamasery high in the Himalayas, located in a remote valley which slows the ageing process of its inhabitants. The 'lost' protagonist, British Consul Hugh Conway, through whose recovered manuscript the experience of a European party—survivors of a plane crash who stumble upon the lamasery—the story is related, is tempted to stay but in the end decides to do the 'decent thing' and escort his young colleague to safety and civilisation. He attempts to return, but whether this is successful is left open. Occasionally, these enchanted places provide much-needed restoration, respite, and reward for weary travellers, as can be seen in The Last Homely House in The Old Wood, the hospitable home of Tom Bombadil and Goldberry Riverdaughter; Rivendell, the House of Elrond; and Lothlórien, realm of Galadriel and Celeborn. Legolas describes the latter as

> "[t]hat is the fairest of all the dwellings of my people. There are no trees like the trees of that land. For in the autumn their leaves fall not, but turn to gold. Not till the spring and the new green opens do they fall, and then the boughs are laden with yellow flowers; and the floor of the wood is golden, and golden is the roof, and its pillars are of silver, for the bark of the trees is smooth and grey."

The incantatory language Legolas uses here, with the alliteration of 'gold; golden; green; grey'; the assonance of the vowels, especially 'o'; and the arboreal symbolism —trees always had a special place in Tolkien's heart— and of course, the echo in the gold and the silver of the Two Trees of the Valinor, Laurelin (the Gold Tree) and Telperion (the Silver Tree), which shone with the light of the Valar until destroyed by Melkor and the primal spider, Ungoliant—all these qualities and associations help to create the enchantment in Legolas' speech, which then transfers onto Lothlórien itself. The forest becomes 'enchanted' through these speech acts, which re-occur throughout *The Lord of the Rings,* as characters talk, sing, and write about its sacred nature in a kind of whispered reverence, seducing the reader with the spell of the natural world.

But what happens when the enchantment is lost?

Lost Paradises

Since the bloody first half of the 20th century, which saw the world decimated by two global conflicts, there has been a growing trend in Fantasy to depict disenchantment, although in the Weberian sense this had been discernible trend since the Enlightenment/Age of Reason, with Matthew Arnold epitomising its tangibility in his mid-Victorian poem, 'Dover Beach' (1867, with early drafts going back to possibly 1849):

> The Sea of Faith
> Was once, too, at the full, and round earth's shore
> Lay like the folds of a bright girdle furled.
> But now I only hear
> Its melancholy, long, withdrawing roar,
> Retreating, to the breath
> Of the night-wind, down the vast edges drear
> And naked shingles of the world.

Of course, this is disenchantment from a Christian perspective, but no less powerful or universal for that, as it captures the zeitgeist (or *lack* of spirit in the age) at the height of the Industrial Revolution. It is interesting to note in the final stanza of Arnold's poem, the world *does* seem like a magical and symbolic place, as he renders it in almost mythic terms, but not one that perhaps would warm the heart of a good Christian soul:

> Ah, love, let us be true
> To one another! for the world, which seems
> To lie before us like a land of dreams,
> So various, so beautiful, so new,
> Hath really neither joy, nor love, nor light,
> Nor certitude, nor peace, nor help for pain;
> And we are here as on a darkling plain
> Swept with confused alarms of struggle and flight,
> Where ignorant armies clash by night.

It feels like we have slipped into a Grimdark novel, and it is not surprising that the subgenre, which emerged from the *Warhammer 40,000* RPG, became the dominant mode of Fantasy at the end of the 20th century and into the early 21st century. Nihilistic, brutal, and Darwinian, Grimdark offers a neoliberalist paradigm of resource-hoarding and survival of the fittest. Magic has bled from the world, or has been deliberately banned. All the tropes of Fantasy are deconstructed, or represented in cynical terms. Major characters die, unexpectedly and often in very violent circumstances. There is no sense of safety. No magnetic north of moral good to navigate by. Everyone is looking out for themselves. This is not a benevolent universe to exist in. And, as such, is entirely resonant with the way the world seemed, post-9/11, and in the ensuing 'War on Terror', which has created a domino effect of heightened geopolitical tension and conflict ever since. This bleak paradigm is seen in the works of George R.R. Martin (who is something of a godfather of the movement, although his diverse oeuvre transcends it), and the self-styled 'Lord Grimdark', Joe Abercrombie, and his emulators. If it could be seen as the 'Punk' phase of Fantasy, blasting out all the flabby clichés of the preceding generation, then perhaps Grimdark served a necessary catharsis, reinvigorating the genre with a fresh energy—a sense of risk and unpredictability, which had been lacking for some time in the formulaic outpourings of Commercial or Commodified Fantasy. A quarter of century on, Grimdark itself has become its own tribute act—its tropes as predictable as the ones it had torn down—and so it is time to move on, and find new ways of imagining Fantasy, ways that are less predicated upon the 'way things are' in the apparent world (which seems like a foreclosure on the imagination, as well on the hopeful, ludic, and joyous spaces of the genre, where *anything* should be possible).

Yet narratives of disenchantment were starting to emerge throughout the 20th century. Hope Mirrlees' sui generis Fantasy novel, *Lud-in-the-Mist* (1926) depicts a petite bourgeois small town that has banned any fruits of fairy from crossing the border they share with the Faerie realm. In its mundanity and conservatism, it is the very epitome of disenchantment. Anything otherworldly is seen as disreputable to the xenophobic citizens (at least to most). Inevitably, contraband is smuggled over and ontological contamination occurs, creating a 'crosshatch'. In *Stardust* (1997) Neil Gaiman offered his own variant on the novel he cites as a major influence.

As awareness of anthropogenic impact on the environment erupted into public consciousness from the early Sixties with the publication of Rachel Carson's *Silent Spring* (1962), elements of environmentalism started to seep into Fantasy. The neo-animism of *The Lord of the Rings*—with its talking trees and nature-loving elves—resonated with the Counterculture, who had adopted the book as something of a Bible (much to the annoyance of the author), which created a feedback loop as the New Wave of Fantasy authors started to weave in such motifs into their imaginary worlds. Michael Moorcock introduced a harder, more cynical edge to Fantasy through his Eternal Champion novels—the time-travelling Jerry Cornelius sequence (*The Final Programme* (1968); *A Cure for Cancer* (1971); *The English Assassin* (1972); *The Condition of Muzak* (1977), plus several short stories, adaptations, and spin-offs) bringing in a psychedelic bricolage of 20th-century aesthetic. Sometimes co-writer of these adventures was fellow *New Worlds* prodigy M. John Harrison, whose sequence of stories set in the bleak, toxic post-industrial landscape of *Viriconium* (1971–1984) steered full into the skid of an anti-pastoral, postlapsarian aesthetic of 'fallen' nature and damaged protagonists:

> The wealth of its people lay entirely in salvage. They possessed no science, but scavenged the deserts of rust that had been originally the industrial complexes of the last of the Afternoon Cultures: and since the largest deposits of metal and machinery and ancient weapons lay in the Great Brown Waste, the Northern Tribes held them.
>
> (The Pastel City, 1971, collected in 2000: 25)

Even more than Tolkien's Middle-Earth, Harrison's Viriconium is a supremely hauntological landscape—a palimpsest of histories, technologies, ruins, and ghosts. It feels like a darker outlier of Mervyn Peake's *The Gormenghast Trilogy* (collected 1992), the hinterlands of which was explored in the third of his original series, *Titus Alone* (1959*)*, which veered unexpectedly into science fiction (SF) as the titular character ventured beyond the suffocating edifice of the crumbling ancestral castle. References to modern technology such as helicopters and telephones were taken out by the editor, but Golden Age SF motifs such as flying cars and death rays remain. Within the space of three novels, Peake has gone from the Gothic to the B-movie, and yet the celebration of the grotesque and the atmosphere of squalid decrepitude that pervades his trilogy help to create some tonal consistency. For Peake, a war artist who had recorded the liberation of Concentration Camps in WW2, the disenchantment was there from the beginning: the rotten core at the heart of humanity and the cruelty behind the façade of civilisation. Life was struggle, torment, and madness—only redeemed by a very dark sense of humour.

Existing in a substantively different universe, Peter S. Beagle's eco-Fantasy *The Last Unicorn* (1968) imagines a unicorn who discovers she is, apparently, the last of her species, until she learns her kind have been driven away by the 'Red Bull'. She goes off in search of them, meeting humans—most of whom are unable to see her horn, for as J.M. Barrie asserts in *Peter Pan*, 'All the world is made of faith, and trust, and pixie dust.' Only those who believe in unicorns can see her true nature. In the stage play version of Barrie's story (*Peter Pan; or, The Boy Who Wouldn't Grow Up*, 1904), which preceded the first full-length book treatment (*Peter Pan in Kensington Gardens*, 1906, although the character first made an appeared in *The Little White Bird,* 1902), it was the audience's belief in fairies that revived Tinkerbell. In *The Last Unicorn,* this Tinkerbell Effect is self-applied by the titular character, whose search for her own kind is a form of self-determination, which many fans of Fantastika could relate to: if others exist like her, then it validates her own existence. The unicorn is captured and made an exhibit of the 'Midnight Carnival', run by a witch, Mommy Fortuna, which comprises illusory and real legendary beasts. The unicorn is assisted in her escape by Schmendrick the Magician. The narrative explores the dialectical tension between illusion and reality. Enchantment remains possible, for those with eyes to see it.

In *The Neverending Story* (German: 1979; English: 1983), which is both a metanarrative and Portal Quest Fantasy, Michael Ende imagines a Secondary World that is threatened with a mass extinction event. A young boy, Bastian Balthazar Bux, discovers a magical tome in an obscure bookshop, which relates a Hero's Journey story-cycle set in the Secondary World of Fantastica. There, a Childlike Empress sickens as her realm is threatened with being overwhelmed by a vortex of destruction known as The Nothing, which relentlessly devours all elements of this fantasy land and turns them into 'lies' in the real world. A young hero called Atreyu is tasked to find a cure, but ultimately the Empress can only be saved by a human child from the real world bestowing upon her a name. This, it turns out, is Bastian, who starts to suspect that the characters in the story are somehow aware of him, but when he is prompted to speak the Empress' name, he refuses. This Refusal of the Call is a classic stage of the Hero's Journey and shows that Bastian is the 'reluctant messiah'. Within the storyworld, characters start to read out from a book also called *The Neverending Story*, this time featuring Bastian himself. Only by entering the story himself can he stop the destruction. He is gifted the talisman, Auryn, which grants wishes to whoever wears it—but with each wish, Bastian loses one of his real-world memories until he even loses his name, and finds himself trapped within Fantastica. This time, it is the turn of Atreyu to name and free Bastian by promising to continue his adventures. Drinking from the Waters of Life, Bastian is restored. Returning to the real world, Bastian is reunited with his ailing father—a Fisher King figure—whom he now restores with the gift of his stories.

In Ende's narrative, it is the act of storytelling itself which creates enchantment, and helps to counter the encroaching disenchantment. In this, and in other narratives of disenchantment it is ultimately acts of belief, of faith, that restore the sickening world, as in Michael Moorcock's *The War Hound and the World's Pain* (1981), the first Ulrich von Beck novel (another iteration of the Eternal Champion), which relocates the Arthurian Wasteland to a Europe ravaged by the Thirty Years War. Here, though, the 'Fisher King' figure is Lucifer himself, who charges von Beck—a mercenary whose soul is damned through his many ignoble actions—with seeking the Holy Grail and healing both the fallen angel and the fallen world of its existential pain. By doing so, the never knowingly altruistic von Beck will save his own soul. However, upon achieving the Grail, von Beck discovers that it will save humanity through rationality and self-emancipation without the help of divine intervention, thus providing the book with the ultimate disenchantment denouement.

Urban Fantasy, a subgenre that emerged in the Eighties in the works of Charles de Lint, Emma Bull, Clive Barker, Neil Gaiman, Terri Windling, and others, was a way of rehabilitating enchantment within a frame of dirty realism, which acknowledged contemporary sociological factors while retaining the possibility of magic, with the added appeal of it being just beyond your doorstep or around the next corner. A successful example of this on the big screen can be found in *The Fisher King* (1991), Terry Gilliam's neo-medievalist Fantasy (written by Richard LaGravenese), which skilfully plays with tropes from Medieval Romance and heroic quest fantasy of the Arthurian tradition. Parry, a teacher, is traumatised by his wife's murder, triggered by a shock-jock radio DJ, Jack Lucas, who inadvertently prompts an unstable man to go on a killing spree. After a psychotic breakdown, which leaves him catatonic, Parry develops medievalist delusions—that the Holy Grail exists, and is being kept in the Manhattan apartment of a dying, wealthy man. Jack Lucas loses his job and ends up working in a video store, a depressive alcoholic. After he rescues a suicidal Jack Lucas from being set on fire by punks, Parry convinces Lucas to assist him on his quest. When Lucas realises who Parry is, and his responsibility for his condition, he sees the fool's errand as a chance at redemption. Gilliam skilfully maintains the realism by presenting any intrusions of Fantasy as the result of Parry's psychosis, thereby allowing audiences to enjoy his trademark visionary flourishes (a red knight galloping down a New York street; busy commuters in Grand Central Station suddenly dancing with one another in harmony), made more poignant and powerful by the frame of verisimilitude. We allow ourselves to be swept along by the magic, while all the time knowing each vision will end, and that the world remains a painful and tragic place. Throughout his post-Python solo film career (notably *Time Bandits*; *Brazil*; *The Adventures of Baron Munchausen*; *Twelve Monkeys*; *The Imaginarium of Doctor Parnassus*; and *The Man Who Killed Don Quixote*), Gilliam has explored this tension between the imaginative and the quotidian. His oeuvre could be seen as sustained one-man resistance to the tides of disenchantment that many who flourished in the Sixties must have felt.

It was only towards the Millennium and new century that the tide from the Otherworld started to turn back.

Paradise Regained

The crisis of disenchantment, first noted by Max Weber in 1918 in the embers of the First World War, when the old world order and the grand narratives of the Victorian age lay in smithereens, reached its zenith in the Thatcherism and Reaganomics of the Eighties. Yet, towards the end of the decade and over the tumultuous years either side of the Millennium, a counter-narrative began to emerge. Foreshadowed by the so-called New Age, which initially seemed like a breath of fresh air in the closed system of hard materialism, but became ultimately rendered meaningless as a commercial marketing category for a particularly anodyne form of spirituality, writing, and music. Yet, within academe a more substantial counter-discourse emerged, heralded by Jane Bennett's *The Enchantment of Modern Life* (2001) and Charles Taylor's *A Secular Age* (2007), among others, and echoed in the many works that have emerged since, exploring the post-Weberian concept of 're-enchantment'. These works collectively attempt to chart what could be called 'Die Neuverzauberung der Welt', the re-enchantment of the world. Jason Crawford critiques these in his essay, 'The Trouble with Re-Enchantment', saying 'the literature of enchantment seems actually to be uncomfortable talking about enchantment' (2020) often qualifying it by wishing to avoid 'magical thinking', the wish to return to a childlike innocence, or any kind of naivete, credulity, or superstition.

And yet Fantasy literature has no problem 'talking about enchantment'.

Contrary to Crawford's concern, there *is* a literature of re-enchantment. The Fantasy sections of bookshops are full of such works. Crawford would argue that '[t]he "alternative" worlds of genre fiction and Mallarmé might offer some solace to the citizens of a disenchanted world, but they don't really change the condition of that world' (2020). I think many fans of Fantasy would disagree. It can fundamentally change one's experience of being in the world, making it not only more bearable, but enjoyable: one can find a community, a purpose, an identity (Lioi, 2016). These are not insignificant things.

The central argument of this book is that Fantasy also enriches our perception of the natural world, and can not only influence our experience of being in it but also inspire conservation, creativity, and future careers, as well as support the tourist economy (with the acknowledgement that the latter can have both positive and negative effects).

Yet Crawford acknowledges the word ('re-enchantment') itself is a slippery, embryonic concept: 'The word itself is still busy being born, a rumour just emerging from our cultural noise' (2020), which only emerged around the Millennium, and has only lately fallen into common usage. 'Disenchantment', on the other hand, carried with it 'affective freight' (2020) which bled into Weber's meaning of the term, and has subsequently become a metonym for modern ills: 'The word disenchantment has, since Weber, come to indicate a condition of historical lateness; a bondage to technologies and ideologies that reduce, exploit, and exhaust; a peculiarly modern form of intellectualized, bureaucratized despair.' Therefore, Crawford argues, 'It's . . . unsurprising that the re-enchantment of the world has become an icon of the postmodern, framed by narratives of loss and recovery' (2020). In our attempts to re-enchant, though, there is often an ersatz replacement of what first generated real enchantment with a simulacrum, a theme-park ride or CGI blockbuster, a Debordian obsession with 'spectacle' (2012 [1967], 1990), at the expense of substance. Yet, a more grounded notion of re-enchantment suggests a re-connection to that which is vital, actual, and embodied: the innate wonder of the world. Rather than the ontological slippage Crawford notes where the real is replaced by the artificial ('estranged by the primary thing by a history of loss'), Fantasy at its best instantiates the world, thus avoiding any degeneration. Even Crawford acknowledges, 'the powers of fantasy can be a conduit for knowledge'.

Fantasy often engages with the Weberian notion of disenchantment—in the narratives of absence, thinning, separation, and seepage mentioned in the previous section—while *also* performing the work of re-enchantment.

This can be seen in texts such as Robert Holdstock's *Mythago Wood Cycle* (1984–2009). The first novel to be published in the series, *Mythago Wood* (1984), starts with the disappearance of the scholar-father, George Huxley, which leads to his two sons (Christian, and his younger brother, Steven) returning in 1947 to the family home, a lodge on the edge of Ryhope: a small section of ancient woodland situated in Herefordshire, on the Borders between England and Wales. They are shocked to discover the woodland's boundaries seem to have *moved*, and it has overwhelmed the lodge. The father's journal is unearthed, and it relates an increasing obsession with the woodland and various 'mythagos', the core concept of Holdstock's cycle.

The author provides several definitions of this slippery term. The first is offered in *Mythago Wood* (1984) by Christian Huxley to his brother, Steven—explaining how in ancient woodland the 'aura' around all living things creates 'a sort of creative field that can interact with our consciousness' (Holdstock, 1984: 49). Paraphrasing their father, Christian says,

> And it's in the unconscious that we carry what he calls the pre-mythago—that's unconscious that we carry what he calls the pre-mythago—that's *myth imago,* the image of the idealized form of a myth creature. The image takes on substance in a natural environment, solid flesh, blood, clothing, and—as you saw—weaponry.
>
> (1984)

The sons venture in the various 'zones' (ash, oak, thorn, et cetera) of Ryhope asynchronously, resulting in dramatically different consequences.

They find themselves sucked into the mythic landscape of Ryhope, which is an extended portal or 'Time Abyss' (Clute and Grant, 1999: 946–947): the further in you go, the bigger (or older) it gets. This trope appears throughout Fantasy fiction; it is 'common to fantasy, uncommon anywhere else' (Clute and Grant, 1999: 586). It crops up in Jorge Luis Borges' *The Aleph* (1945) as an object that contains the whole universe, in the year the first atomic bombs were used in warfare and, as such, is not surprisingly a lingering device of the Atomic Age. However, the sense of expansive interiority and dilatory disjuncture are common descriptors in tales of 'hollow hills' and other portals to the Otherworld.

It is not until Holdstock's last published novel, *Avilion* (2009), that Steven and Guiwenneth are reunited. They settle down in the valley in an old Roman villa (an architectural mythago conjured from Steven's memory) and have a couple of children, Jack and Yssobel. *Avilion* resumes the story at the point when this idyll is shattered. Guiwenneth has disappeared, and Yssobel has gone to find her—heading inwards to Avilion/Lavondyss. Jack (half-human, half-mythago like his sister—both 'red' and 'green') journeys outwards to the edge of Ryhope—desperate to see the world of his father, the Lodge where it all started, the village of Shadoxhurst, and hopefully find some clues that will help in the search. By the time Jack reaches 'the fields we know', it is the present day (the original disappearance of the three Huxley men—George, Steven, and Christian—occurred in the Forties). As all travellers of Faerie discover, time runs differently in each world. In an intertextual touch, Steven has quested Jack with finding his old copy of H.G. Wells' *The Time Machine* (1895) in the Lodge. Jack's presence there dislodges the ghost/mythago of his grandfather, George Huxley—and we have some insight into the original chain of events.

Overall, Ryhope Wood acts as both a portal to otherworlds (the deep past; Faerie) and a place of initiation, transformation, and generation, interfacing in a mysterious way with anyone foolhardy enough to penetrate its fastness to create physical manifestations of folklore and ancestral memory.

The novel's conversation with genre is worth noting. As Raelke Grimmer concluded in *Gothic Nature*,

> Far from being prescriptive, actively engaging with genre and using such conventions as tools in writing aids the creative process. It is by thinking about genre and crossing genre boundaries that new genres emerge in a renegotiation of the changing social contexts in which writers create their works.
>
> (Grimmer, 2016)

I would argue that Holdstock's cycle does this, although it is conventionally categorised as Fantasy, and has received accolades as such (winning the World Fantasy Award in 1984). For some, such a designation stigmatises it beyond the pale of academic discourse, for, as Le Guin wryly puts it, as a literary tradition, Fantasy is 'distinguished in college English departments mainly by being ignored' (2004: 42), albeit with notable exceptions (e.g. the sister genres of Gothic and SF). Holdstock's fiction has, in singular fashion, followed 'a deep current in one direction—towards rejoining the "ocean of story", fantasy', in Le Guin's sense of 'things not actually present' (2004). And yet there is a hybrid quality to Holdstock's cycle as it imports science fictional and faux-scholarly textual elements into the prose. Events are given a pseudo-scientific frame and are layered with psychological ambiguity. Through this stylistic crosshatch Holdstock the scientist, trained as a medical zoologist, debates with Holdstock the genre author.

Running throughout the *Mythago Wood Cycle,* but especially tangible in *Mythago Wood* (1984) and *Lavondyss* (1988), is a narrative of re-enchantment. The two brothers return to Ryhope in the bleak, rationed years after the Second World War. Steven Huxley has returned from active service after recovering from his wounds; a family friend, Harry Keeton, is a burn-scarred RAF veteran. Like the two traumatised veterans in J.L. Carr's *A Month in the Country* (1980), there is a lingering legacy of Die Entzauberung der Welt—reality has been brutalised; hearts and minds cauterised by conflict, tragedy, and atrocity—which perhaps only the convalescent and healing powers of nature can restore. The protagonists of Holdstock's books are often wounded or damaged in some way, and it is tempting to risk reading both biographically and autobiographically into his narratives, for Holdstock's grandfather narrowly survived the Trenches of the First World War—an experience he returns to directly in poetry and indirectly in prose (interview with Kincaid, 2011)—and the author suffered mental health problems. Clearly, Holdstock found re-enchantment in nature (and in the evocation of rural England found in Ralph Vaughan Williams' 'Lark Ascending'), and this is conveyed through his complex tangle of narratives—even if Ryhope is a place of testing, initiation, and transformation, the woodland itself offers a kind of refugium of the imagination from the 'too much reality' of the world. Re-enchantment occurs in the frisson of language on the page, and by the evocation of magical realities, especially when they are given an earthy, tangible quality as they are in Holdstock's oeuvre. As Michael Moorcock himself has noted, 'No other author has so successfully captured the magic of the wildwood' (1984). There is a visceral, animalistic quality to Holdstock's descriptions of his mythagos, which makes them seem utterly believable. And in terms of the current book's central argument, the reading of *Mythago Wood*

(1984) completely transformed *this* reader's perception of woodland. Sylvan settings had long been places of refuge and dreaming, but entering one was never the same again, for Holdstock created what bryologist Robin Wall Kimmerer has called a 'grammar of animacy' (2017). Not only did he make the wood seem alive in a preternatural way—full of ontological potentiality—Holdstock dramatised the act of enchantment, alternatively called the creative process, itself, as I have explored elsewhere (2018).

As this brief foray into Epic Fantasy has shown, setting is never merely a backdrop, but a key part of the narrative. It can have sentience, agency, even character. It can also be seen as an extended metaphor of the journey through life itself, as was certainly intended in William Langland's *Piers Plowman*, Geoffrey Chaucer's *Canterbury Tales*, John Bunyan's *The Pilgrim's Progress*, Wu Cheng'en's *Journey to the West*, and other early works. The scope of the epic provides a suitably large canvas to depict a sense of the intense scope of a single life (or of multiple intersecting lives). It offers, in effect, a spatial metaphor of a temporal existence: a personal timeline becomes an itinerary, a route marked on a map with its landmarks, its nodes of memory and songlines of significance. Yet, sometimes the 'landscape' is closer to home, as we'll explore in the next chapter.

4 Urban Fantasy

The Re-Enchantment of the City

More than a century on from the Industrial Revolution (which created a significant migration of the population in industrialised countries from the rural to the urban) and two world wars the experience for the bulk of the population in the Global North was predominantly urban (Alvarez-Palau et al., 2020). Therefore, it is perhaps understandable that Fantasy should reflect that. Even while there remained a dominant strand of Fantasy that offered 'escapism' into other worlds with a prelapsarian aesthetic, or medievalist reversion of technology and societal progress, a new strand began to emerge in the 1980s which relocated Fantasy to the cities, and offered a darker, more diverse, and more worldly recalibration of familiar tropes. These new stories (eventually identified and marketed as 'Urban Fantasy') were pioneered and fashioned by writers such as Emma Bull, Charles de Lint, Elizabeth Hand, Anne Rice, Neil Gaiman, and Terri Windling. These writers helped forge a new aesthetic for Fantasy, one that was copied in countless films and TV dramas, such as *Time Bandits* (1981); *Brazil* (1985); *Highlander* (1986); *Wings of Desire* (1987); *The Navigator: a medieval odyssey* (1988); *The Fisher King* (1991); *Buffy the Vampire Slayer* (1997–2003); *What Dreams May Come* (1998); *The 10th Kingdom* (2000); *Doctor Who* (new series 2005-); *Supernatural* (2005–2020); *Grimm* (2011–2017); *and Stranger Things* (2016–2025), to name a selection.

Arguably, Urban Fantasy had long been established in superhero comic books since the 1930s—larger-than-life characters with supernatural powers, often drawn directly from folklore, myth, and legend, clash against the backdrop of Main Street, USA. Metropolis and Gotham in the DC comic books, and later New York in the Marvel Universe, became well-established as places of Intrusion Fantasy—whether it was Kal-El adopting the alter-ego of Clark Kent to 'pass' as a human while working as a reporter at *The Daily Planet*; Bruce Wayne tackling the demi-monde which threaten to overwhelm Gotham in the guise of Batman; the Mole Man erupting from beneath Manhattan thwarted by the Fantastic Four;

DOI: 10.4324/9781003463665-5

Galactus from the stars threatening to destroy the entire world; or the Sinister Six of *Spiderman* being continually defeated by the teenager Peter Parker, bitten by a radioactive spider and forced to juggle the demands of his job and social life with fighting crime.

Yet even before these parallel iterations of Urban Fantasy in popular culture, the re-enchantment of the city had begun even in the depths of the Late Victorian Age, as evidenced in the proto-texts of psychogeography.

The Psychogeographical City

The parallel traditions in Psychogeography of the Robinsonade and the Flâneur (Coverley, 2018) jointly helped in the early re-enchantment of the city. Psychogeography, in its broadest sense, has a long and fascinating tradition. Although Debord claimed and colonised the term in post-war France (first in the Letterist pamphlet *Potlatch*, 1954; and then in numerous pronouncements via its evolution, the Situationist International, from 1957), there are many antecedents, influences, and developments. In two distinctive traditions, one based in London (the Robinsonade) and the other in Paris (the Flâneur), leys of affinity can be gleaned: although as with Alfred Watkin's 1922 notion of the 'ley', how much is geographical serendipity, geomantic intentionality, or the projection and preoccupations of the viewer is hard to say. In hindsight, viewed from the hill of the here-and-now, there seems to be a parallax *movement* emerging autocthonically from the labyrinths of London and Paris. Debord coined the term and defined it at length in his contributions to *Potlatch* and later publications. In brief, he described it as '[t]he study of the specific effects of the geographical environment, consciously organised or not, on the emotions and behaviour of individuals' (Debord in Coverley, 2018: 10). Many have tried to define this notoriously slippery term since. Robert Macfarlane defined psychogeography as 'The exploration of cities and other landscapes by means of drift, play and randomly motivated walking, encouraging a re-imagining of familiar terrain' (Macfarlane, 2019). But in terms of our focus here—Urban Fantasy—there are proto-texts that seem to foreshadow that in the Late Victorian and Early 20th centuries.

Early texts that paved the way for psychogeography include *The Man in the Crowd* (1840), a London-based short story by Edgar Allen Poe written at a time when the English capitol was the largest city in the world; *The Strange Case of Dr Jekyll and Mr Hyde* (1886) by Robert Louis Stevenson; and *The London Adventure or The Art of Wandering* (1924) by Arthur Machen. These texts not only showed a different side to the city, often revealing its dark underbelly, they also showed a different way of navigating the city—of moving through urban space, which in itself can be an act of enchantment: a different way of seeing the world, one of simultaneity with a deep awareness of the palimpsest of history and where interiority and exteriority blend into a kind of pathetic fallacy (a repurposing of the lyricism of the Romantics, which foregrounded personal sensibilities over objective empiricism). This way of seeing the visionary artist and poet, William Blake called 'Fourfold Vision', and one could argue that *he* is in fact the Godfather of Psychogeography because of the way he re-enchanted his beloved native London, at the age of 7 or 8 seeing angels in the trees in Peckham Rye: 'A tree filled with angels, bright angelic wings bespangling every bough like stars.' For Blake, everything that lived was Holy, and the world was immanent with metaphor: the backdrop for the vast dramas of his cosmology. From the land arose 'emanations', genius loci such as the spirit of Albion, and to these he gave voice. To depict these he pioneered a way for text and image to be printed at the same time on exquisitely engraved copper plates, what he called the 'Infernal Method'. To modern eyes, these coloured chapbooks, which he self-printed and self-published, look like prototype comic books or graphic novels. If he were alive now, perhaps Blake would be using the platforms of digital streaming media such as Netflix or Amazon Prime to share his cosmic sagas of epic struggles. In one of his poems found in what is known as 'The Pickering Manuscript' (1969 [1803]: 429–430), Blake describes 'another London'. In the poem, Blake describes being caught by a fairy maiden and placed within a crystal cabinet, which allows him to perceive another version of his urban environment:

> Another England there I saw
> Another London with its Tower,
> Another Thames and other hills,
> And another pleasant Surrey bower.

There he also beholds the maiden as a more fundamentally beautiful version of herself (sister perhaps of Coleridge's 'damsel with a dulcimer', or Keats' 'la belle dame sans merci'), but when he went to 'seize' this elusive muse figure, the results are calamitous:

> I strove to seize the inmost form
> With ardour fierce and hands of flame,
> But burst the Crystal Cabinet,
> And like a weeping Babe became—

This could be, in microcosm, the plot of an Urban Fantasy novel, in which the protagonist is haunted by an otherworldly vision of beauty, one that transports them to another world (or one that motivates them to pursue their object of desire across the border between worlds), but one that cannot be 'won', or sustained. As in many folk tales, for any who hear or see Faerie, there is often a terrible price: they become fey, and do not last long in this world.

Another example of perceptual defamiliarisation akin to Blake's Fourfold Vision, albeit with at least one foot in this world, is the literary quality of the Chestertonian 'mooreeffocian'. First cited in John Forster's *Life of Dickens* (1872), it signifies a different way of seeing the everyday, as described by Dickens himself, relating an epiphany in a London coffee-shop:

> In the door there was an oval glass plate, with COFFEE-ROOM painted on it, addressed towards the street. If I ever find myself in a very different kind of coffee-room now, but where there is such an inscription on glass, and read it backward on the wrong side MOOREEFFOC (as I often used to do then . . .) a shock goes through my blood.

Chesterton then referred to it as 'mooreeffocian' in his *Charles Dickens: a critical study* (1906), describing the effect as 'the queerness of things that have become trite, when they are seen suddenly from a new angle'. Although Dickens deployed his own version of this, especially in his short comic stories, it was Chesterton who took the concept and ran with it, deploying it to develop his distinctive brand of Fantasy, notably in his metaphysical detective novel *The Man Who Was Thursday* (1908), which depicts an alternative vision of London hiding in plain sight where a nefarious organisation led by the monstrous nemesis Sunday seeks to wreak havoc upon the world. Into this demi-monde, Gabriel Syme, a double-agent recruited by an anti-anarchist department of Scotland Yard, inveigles himself—becoming recruited as the anarchist conspirators' new 'Thursday', which allows him access to the inner executive. This executive is described as meeting in the backrooms of restaurants, only a door away from the everyday, and their main organisation meets underground in secret basements—epitomising the very subconscious 'shadow' of society and the city.

And so the city became transformed, through such works, into a place of imaginative possibility. However, this is not sufficient to qualify as the 'Ecological Imaginary', as nature is a mere backdrop in such narratives, relegated to civic parks, the river, and the occasional bout of bad weather. For that we need to look at some sui generis works.

Richard Jefferies' singular post-apocalyptic novel, *After London: or Wild England* (1885), describes the rewilding of England following the appearance of a mysterious dark orb (whose appearance is never explained) which effects the tides and causes mass flooding. Packs of feral animals roam the land. After 150 years, that which is not covered by water has been reforested. Civilisation has reverted to a Medieval level of technology and society in what is largely a future primitive scenario. There is a slight plot around a disinherited knight, Felix Aquila, attempting to reclaim his lands and win his sweetheart, but the first part of the novel is a purely descriptive extrapolation upon the effects of the flooding and rewilding of England. Although this does not make for a very effective novel, dramatically (going even further than Thomas Hardy, who spent the first chapter of *The Return of the Native*, 1878, describing the main setting of Edgon Heath, a fictionalised of the Black Heath that once stretched from Dorchester to Poole), this bold re-imagining (preceding the current 'rewilding' movement by over a century) is an inarguable, if extreme, example of the Ecological Imaginary—one that shifts the emphasis to the non-anthropocentric. It is in the later sections of the novel that we encounter an early iteration of Urban Fantasy, when Felix paddles into the heart of darkness of a completely destroyed London—now rendered into a toxic wasteland. The former capitol is all but buried beneath a deep black sediment of countless bodies and accumulated corruption, echoing the Biblical judgement on Sodom and Gomorrah and foreshadowing the nightmarish images of firebombed European cities, and Hiroshima and Nagasaki decimated by the atomic bomb. Foreshadowing the eerie vistas of England subdued by the invasive species of the 'red weed'—part of the biological terraforming by the conquering Martians in H.G. Wells' *The War of the Worlds* (1898)—Jefferies vision does not offer solace, but a certain misanthropic schadenfreude, a rueful warning of what may (deservedly, in his mind) befall humankind if it continues down its cult of (unsustainable) Progress. Nature is not re-enchanted in Jefferies' apocalyptic vision, but re-instated.

A more positive alternative can be found in Arthur Machen's short story, *N* (1935). This is a 'gentleman's club' story, in which three friends, ensconced comfortably around a merry hearth with glasses of steaming punch, reminisce on the London they knew. At one point the conversation drifts to Stoke Newington, where one of their fraternity relates three anecdotes that allude to a secret arcadian inner-landscape glimpsed in the midst of a very urbanised corner of the capitol. Looking through a bedroom window, the narrator in his anecdote describes this unexpected vista:

> 'For a moment, my heart stood still, and I gasped for breath. Before me, in place of the familiar structures, there was disclosed a panorama of unearthly, of astounding beauty. In deep dells, bowered by overhanging trees, there bloomed flowers such as only dreams can show; such deep purples that yet seemed to glow like precious stones with a hidden but ever-present radiance, roses whose hues outshone any that are to be seen in our gardens, tall lilies alive with light, and blossoms that were as beaten gold. I saw well-shaded walks that went down to green hollows bordered with thyme; and here and there the grassy eminence above,

> and well before, were crowned with architecture of fantastic and unaccustomed beauty, which seemed to speak of fairyland itself.'
>
> (Machen, in Miller, 2020: 222–223)

The protagonist goes in such of this 'phantasmagoria', to no avail. Like the Arcadia described in *La Grande Meaulnes* (1913), Alain-Fournier's only novel, written before he was killed in the First World War, it is a vision once-glimpsed and never to be regained.

The psychogeographical tendency in fiction was continued by the Inkling, Charles Williams (in his occult novels such as *War in Heaven*), and then later Peter Ackroyd and Iain Sinclair: the British high priests of psychogeography, who, through both of their oeuvres have done more than anyone to re-enchant their beloved London—peeling back the layers of the city to reveal its incredibly rich history. Alan Moore has tapped into this rich seam also, first in his 'sonic ritual', *The Highbury Working: a Beat Séance* (2000), and most recently in his project *Long London* series of novels (which began with *The Great When* in 2024).

After this deríve into the psychogeographical city, let us now turn to the edgelands.

Edgelands

The British poets Paul Farley and Michael Symmons Roberts, in their book *Edgelands* (2011) sought to celebrate the 'zones of inattention' they noticed in the liminal places of urban landscapes: brownfield sites, motorway infrastructure, abandoned factories, scruffy housing estates, and boarded-up shopping malls—the architecture of the banal. Iain Sinclair performed his own psychogeographical deríve of the M25 in *London Orbital* (2003), and in *Scarp* (2013) Nick Papadimitriou explored the 14-mile ridge in North London near his home in an endeavour he called 'Deep Topography': another praise-song to the capitol's (un)loved landscapes. Patrick Keiller's *Robinson in Space* (1997) is a cinematic meditation on such hinterlands, using the device of its fictional flâneur to explore the psychic spaces of liminal London. The naturalist Richard Mabey, in his book *Weeds* (2010), drew a botanist's gaze to what flourishes in these interzones, in Victor Hugo's 'bastard countryside'. Yet other kinds of biodiversity can be found here, too.

Thresholds are hot zones of creativity—what the American poet Gary Snyder calls in *The Practice of the Wild* (1990), a 'burning edge . . . a frazzle'. They are places of creative tension, where difference is encountered, negotiated, and interpenetrated. Such places have long been a source of rich inspiration for writers of Fantastika, such as William Hope Hodgson's

The House on the Borderlands (1908), which imagines a remote house in rural Ireland situated on the edge of the otherworld: an infernal abyss from which swine-like creatures emerge to terrify the unfortunate inhabitants (echoing the Brythonic tradition of pigs being a 'gift' from Arawn, Lord of Annwn, the Welsh Hades). More relevant to our focus here on Urban Fantasy is Hope Mirrlees' singular Fantasy novel, *Lud-in-the-Mist* (1926), which imagines a petite bourgeois town on the edge of Faerie (at the confluence of the Dapple and the Dawl in the fictional country of Dorimare), where fairy fruit is contraband and any cultural cross-contamination from their neighbours is highly frowned upon: an embargo which is inevitably broken by no less than the mayor himself, Nathaniel Chanticleer, who succumbs to the temptation of the fruit, like the sisters in Rossetti's 'Goblin Market'. His wish is to save his son from the bad influence of Faerie, but he himself falls victim to its charms.

The riparian setting of Mirrlees' Fantasy of Manners is perhaps a conscious echo of the setting of the Scottish Border Ballad, 'Tam Lin', Carterhaugh, which is situated at the convergence of two rivers, the Yarrow and the Ettrick, a place called 'The Meetings' on The Duke of Buccleuch's estate in Selkirk. Having been warned to stay away from Tam Lin's well, Janet o' Carterhaugh is compelled to go there to try her luck, similarly breaking the taboo around Faerie like the Mayor of Lud-in-the-Mist. She 'pulls a rose' and summons the fairy knight—only to discover he is a mortal trapped by the Queen of Elfland, no less. But he informs her of a way to free him—on Hallowe'en when the Fairy Rade passes by. Janet manages to pull the ensorcelled Tam Lin down from his horse, and hold onto him as he undergoes various zoomorphic transformations, while the Queen balefully watches on, uttering a rueful curse as the lovers depart. This story has been explored in several novels (Dean, 1991; Maas, 2015; Manwaring, 2018).

Deeply inspired by Mirrlees' novel, Neil Gaiman set out to write his own variation on her novel, *Stardust* (1999), composing the initial draft in purple ink in a large notebook. Set in the village of Wall, named after the long-standing boundary which demarcates the border between the mortal and Faerie realms, a conservative Lud-in-the Mist type settlement where a disreputable magical market is held every nine years, it relates the adventure of Tristran Thorne (who is of half-human, half-faerie blood), who, in wishing to win the heart of a local beauty, Victoria, vows to bring her back a fallen star from over the border. This fallen star turns out to be the fairy princess Yvaine, whom he falls in love with while escorting her back to his home village. Yvaine is sought after by the Lords of Stormhold, and many picaresque adventures ensue in the manner of a classic Romance.

Closer to home, Sylvia Townsend Warner was an American writer who settled in the Dorset village of Chaldon Herring as part of an artistic coterie, along with her partner, the artist Valentine Ackland. Warner wrote

Kingdoms of Elfin (1977), a wry and whimsical set of Fantasy vignettes in the manner of Lord Dunsany's *The Gods of Pegāna* (1905) and *The King of Elfland's Daughter* (1924). And yet it is her Fantasy novel set in this world, *Lolly Willowes* (1926), that concerns us primarily here. The titular character is a middle-aged spinster who decides to leave London to escape the controlling influence of her relatives, moving to the village of Great Mop (a fictional village set in the Chilterns). While out walking, she decides to make a pact with Satan out of frustration with all the societal pressures forced upon women. Afterwards, she befriends a cat, which becomes her familiar, calling it Vinegar; and her landlady invites her along to a witches' sabbat, where she finds the empowered female community she had been seeking all along. Lolly decides to take up witchcraft in defiance of her family, and the stifling conventions of early 20th-century society. *Lolly Willowes* could perhaps be more accurately called Rural Fantasy, but it does migrate from the city to the countryside, although it is the latter that is the place of imaginative possibilities, where other ways of being in the world are afforded away from the yoke of city and society.

A novel that definitively locates the Fantastic within the urban is Mikhail Bulgakov's *The Master and Margarita* (1st English translation, 1967 [1924–1940]), an excoriating satire of Communist Russia rendered as an absurdist comedy with Magic Realist elements. As the Devil runs amok in Moscow (the enigmatic and dangerous magician Woland, and his fiendish entourage Koroviev, Azazello and Behemoth—the latter, a giant black cat who walks on two legs), he uses the hellraising as a way of challenging the status quo and deconstructing the whole system of bureaucratic madness, corruption, and superficiality. The real black magic is enacted by the State, and perpetuated by every citizen in countless petty ways. The main target of his satanic barbs is the literary establishment (especially the theatrical scene), which made his life hell—the critics get a particular roasting. Bulgakov mocks the writerly cliques with ridiculous acronyms—popular at the time—and those that are 'in' and 'out' of favour. At one point, Koroviev cries, 'A writer is defined by his writing, not his membership card!' Bulgakov's devilish clowns gallavant around Moscow like the Marx Brothers, continually being 'inappropriate' and gatecrashing elite clubs—and as Groucho once said: 'I don't care to belong to a club that accepts people like me as members.' Bulgakov delights in the polysyllabic names of the Muscovite literati. A typical example is 'Arkady Appolonovich Sempleyarov, chairman of the Acoustics Commission'. Bulgakov rattles out these long-winded nomenclatures in an almost incantatory way. Nothing is simple or as it seems. It is a city of masks and walls and mirrors—and the Devil delights in pulling away the veil to reveal the ugly truth. The novel has a metanarrative element—a subplot concerning a writer with a semi-burnt magnum opus about Pontius Pilate and the execution of a certain 'religious zealot'. The plot switches between this fully imagined recreation of the Crucifixion (and its immediate aftermath) and the dramatic present of 1929 Moscow. The fun is to be had in this 'present day'—chiefly the intrusion of gods and demons into a secular, harshly atheist Russia. Things get increasingly apocalyptic towards the end—and Bulgakov at times depicts scenes reminiscent of the apocalyptic sublime art of John Martin, and also the works of some Symbolist painters. Yet all is told with a devilish wit—casting contemporary 'end of the world' forebodings into an ironic light. Bulgakov died before finishing editing this censored novel—written in secrecy, burnt, and banned for decades. Created from the two surviving texts, it is an unparalleled swansong from one of Russia's most important writers: a radical, activistic act of authorship. It is undisputably Urban Fantasy, and yet it is perhaps less useful to us as an example of the Ecological Imaginary, because Bulgakov is not so interested in nature here, except in the sense of 'human nature'. For a more full-bloodied engagement with the natural world, one must turn to his short story, *The Heart of a Dog* (1968 [1925]), which imagines the comical transference of qualities between a man and a canine (again, with a satirical agenda—in this case, a savage criticism of Bolshevism).

For a more sustained deep dive into the Ecological Imaginary in urban spaces, we need to turn to a fellow Russian of Bulgakov's, the film-maker Andrei Tarkovsky, whose SF film *Stalker* (1979), hauntingly depicts an eerie radioactive wasteland. A guide simply known as 'Stalker' is hired to lead a professor and a writer into this 'Zone', in search of a room that grants wishes. The Zone affects them physically and mentally, sending them into themselves. They encounter strangely prescient dogs and edgelands of praeternatural beauty and ugliness. It is an area where the accepted laws of physics break down. Tarkovsky's film is loosely based upon the SF novel, *Roadside Picnic* (1972) by Arkady and Boris Strugatsky, which imagines the aftermath of an extraterrestrial encounter called the 'Visitation', which leaves the area around it (the 'Zone') affected in unusual ways. Clearly drawing upon the legacy of nuclear tests, and eerily foreshadowing the impact of the Chernobyl nuclear accident of 1982 (literalised in the 2007 unofficial computer game sequel, S.T.A.L.K.E.R: The Shadow of Chernobyl), both *Roadside Picnic* and *Stalker* re-imagine the edgelands of the Atomic Age. In the novel, the Visitation impacted six zones around the world, suggesting similar situations play out across the globe. In his *Southern Reach* series (2014–2024: *Annihilation; Authority; Acceptance; Absolution*), Jeff VanderMeer imagines a near identical scenario, relocated to the Florida Everglades, renamed 'Area X': a place of ontological slippage and metamorphosis that destabilises notions of selfhood, sentience, and civilisation.

In Terri Windling's shared world anthology, *Bordertown* (1986), the setting is the titular settlement on the borders of Elfland, where 'magic meets rock 'n' roll'—a transactional place of bartering, conflict, alliances, feuds, love affairs, and cross-pollination. Windling invited a select group of her fellow Urban Fantasy writers to extend the dramatic possibilities of this classic interzone (as depicted in the Carol Reed film noir classic, *The Third Man*, and in many narratives since), and it continued in several anthologies. Here, the 'biodiversity' is supernatural, with many folkloric figures co-mingling with humans—a conscious metaphor of the multiculturalism that was at that time on the ascent: something that was encouraged and celebrated, rather than being perceived as an existential threat and weaponised by the Alt Right post-9/11.

For a while, in the late 80s and 90s, as the Berlin Wall came down, Glasnost thawed the Cold War, 'New Age' titles filled the bookshops and World Music took off, the borders of the imagination remained open, and other ways of being in the world managed to slip through. But as the new xenophobia of Islamophobia and manufactured Culture Wars took over, a kind of islandisation occurred and surviving pockets of Fantasy found themselves separated and sealed off from the consensus reality: it is to these 'polders' we turn next.

Polders

John Clute defines 'polders' as

> enclaves of toughened reality demarcated by boundaries (thresholds) from the surrounding world . . . an active microcosm, armed against the potential wrongness of that which surrounds it, an anachronism consciously opposed to wrong time. Polders change only when they are being devoured from without.
>
> (1999)

Examples of polders first appear in myth and legend, and then manifest in popular culture. They include Atlantis, Ys, Lyonesse, Tir nan Og, Brigadoon, Shangri-La (as depicted in James Hilton's 1933 novel, *Lost Horizons*), Gormenghast (a medievalist city-state bound by arcane and archaic traditions in a science fictional world), Aldous Huxley's doomed utopia, *Island* (1962), Wakanda, and Hogwarts, to name a few.

John Crowley's *Little, Big* (1981) depicts a house called Edgewood—constructed by the protagonist, architect John Drinkwater, and situated in the north of New York State. Composed of many styles, it echoes the historical Winchester House in San Jose, CA, with its many rooms and extensions (although less tragically haunted by the legacy of the blood-drenched rifle dynasty), acting as a portal to the world-within-the-world. As with the TARDIS, it is bigger on the inside than outside, and is a literalised

Foucauldian heterotopia. It is a place where human and fairy, history and folklore, text and the oral tradition, co-mingle, and the ontological imaginary is expanded.

Edgewood is echoed in the DC title first published in 1982, *Night Force*, which features Baron Winter, who is bound to a house of similar properties, Wintersgate Manor, which is situated at a crossroads of time and space, allowing him to assemble and despatch an eclectic team of agents drawn from history to fight the nefarious forces of evil which threaten to overwhelm the world. Conceived by Horror writer Marv Wolfman with long-time artist collaborator Gene Colan, it was intended to be a format for a genre-fluid anthology series, which allowed the team to draw upon the full 'canon' of Horror and Horror-adjacent characters and creatures: vampires, werewolves, zombies, witches, monsters, and the like. Cancelled after only fourteen issues, due to the collapse of the comic book Horror market, *Night Force's* cast of intertextual characters (such as Vanessa Van Helsing, the granddaughter of the famous doctor from *Dracula*) could be seen as a foreshadowing of Moore and O'Neil's *League of Extraordinary Gentlemen* (1999–2019). Unlike the latter, Wolfman and Colan's Edgewood was a polder that could not withstand the forces of commerce.

A more successful and sustained polder can be found in Charles de Lint Tamson House. Situated in Ottawa, it is the 'size of a city block', and provides temporary or permanent residence for a wide cast of characters: misfits, mavericks, wanderers, and magicians. Like The House on the Borderland, Edgewood, and Wintersgate Manor, Tamson House is located at the fault-line between worlds: 'Built at a point where the leylines meet, upon land that was once a sacred site, it is the gateway to a spirit-world where Celtic and Native American magicks mingle and leak into our own' (1992). Presided over by Sara Kendell and her uncle, Jamie Tamson/Tams—occultist and antiques dealer—it first makes an appearance in de Lint's Urban Fantasy, *Moonheart* (1984). The publication of this had such an impact on readers that some were compelled to track down and identify Tamson House's precise location, as John O'Neill points out in blog on the book: 'It was enough of a local sensation that readers could frequently be spotted in the Glebe in downtown Ottawa, copies of *Moonheart* in hand, pacing out the geography of Tamson House like tourists circling the Eiffel Tower.' De Lint extended his tales of Tamson House in *Yarrow* (1986), *Greenmantle* (1988), and *Spiritwalk* (1992). What qualifies Tamson House as a key text in the urban Ecological Imaginary is the fact that it encloses a garden in which the Fantastic thrives:

> In the overgrown garden of Tamson House, a Coyote Man waits, green children walk, and music rises to greet the moon. From the garden, a

> vast and primal wood is just one spirit-step away . . . and in that wood is something that threatens the very existence of Tamson House, and all who dwell within.
>
> (*Spiritwalk* blurb, 1992)

This vista, afforded the fortunate guests of Tamson House, is akin to that Glanville's guest looks upon in Arthur Machen's *N*, and also the house that Christopher Reeve's character, Richard Collier, walks into from the past in the film *Somewhere in Time* (1980). Woodland is a constant presence in de Lint's Urban Fantasy—perhaps not surprisingly for a Canadian author with a deep sense of the indigenous, storied landscape of First Nations people; in many ways, de Lint is the New World counterpoint to Holdstock—both of whom evoked the magic of the wildwood better than anyone (yet). It is interesting to note that their polders (Ryhope Wood; Tamson House) have a strong arboreal quality and are depicted as sites of active resistance to modernity (two world wars in Holdstock's cycle; gang crime in de Lint's), and both came out of a time when there was a growing sense of our impact upon the environment, such as the deforestation of the Amazon rainforest (perhaps the most precious polder of all). Indeed, Holdstock went on to write the novelisation of John Boorman's eco-narrative film, *The Emerald Forest* (1985).

It is arguable that all coastal cities are, in a very real sense, polders (a term related to the reclamation of low-lying land) and subsequently face severe flooding due to the projected rises in sea level, along with the increasing frequency and severity of storms caused by anthropogenic Climate Change. Currently, 11% of the world's population lives in coastal areas. According to the United Nations, coastal flooding will impact over 70 million people. By living outside of the 'doughnut' of sustainable growth (Raworth, 2017), *all* cities are effectively polders with a limited timespan until they will implode and collapse. Nature will not be benign presence in these instances: floods, wildfires, storms, pandemics, and deadly heatwaves will threaten all human life (Wallace-Wells, 2019; Thunberg, 2022; IPCC, 2024)—an existential threat explored by writers such as Cormac McCarthy in *The Road* (2006), Kim Stanley Robinson in *The Ministry for the Future* (2021); Stephen Markley in *The Deluge* (2023); and Tim Winton in *Juice* (2024), to name a small selection of the growing subgenre of ecofiction (Manwaring, 2024).

Faced with such a nightmarish prospect that only the most radical systemic change will ameliorate and address, it is understandable why many seek refuge in the polders of Fantasy. Yet sometimes the only way to cope with these existential threats is to sidestep them completely, and that is what we turn to next: wainscots.

Wainscots

John Clute describes what he calls 'wainscot societies': 'invisible or undetected societies living in the interstices of the dominant world—normally but not necessarily human' (1999), after the idea that mice, rats, and other more-than-human critters can live co-terminously with humans behind the skirting boards of our homes. In Fantasy this idea has created a small, but resilient tradition of wainscot fantasies, perhaps the most obviously environmental and influential of which is Mary Norton's *The Borrowers* series (1952–1982), which imagines the tiny human Clock family living in such interstitial domestic spaces (behind a clock), and surviving via scavenging and 'borrowing' from the Big People, the 'human beans'. This diminutive race even borrow their names from the objects they find around them. Norton's books have been adapted several times, including in anime form (*Arrietty,* Studio Ghibli, 2010). This Lilliputian trope was first given literary form by Jonathan Swift in his mock-travelogue, *Gulliver's Travels* (1726), but permeated popular culture throughout the latter half of the 20th century, including *Fantastic Voyage* (1966), *Land of the Giants* (1968–1970), *The Micronauts* (1979–1986), *Honey, I Shrunk the Kids* (1989), and the BBC children's TV series, *The Wombles* (1968–1976), the latter describing a race of conscientious critters imagined by their creator, Elisabeth Beresford, to resemble teddy bears with hedgehog-like faces living beneath Wimbledon Common (where a Boxing Day walk with her children gave her the idea after her daughter, Kate, repeatedly mispronounced the park as 'Wombledon Common') continuing the environmental message with their motto: 'Make Good Use of Bad Rubbish'.

Tove Jansson's Moominverse is revealed to be a wainscot in the final book she authored in her beloved series, *Moominvalley*, in November 1970. Up until this point, it seemed like the Moomins and their fellow beings lived in an alternate world—one where there was sorrow and hardship, but no evidence of human civilisation. In this last book, where the Moomin family is tellingly absent, this realm is shown to intersect this world, through the presence of Toft, a small, shy orphan boy who can conjure things into being through his imagination (thought of as a cypher for Tove herself). He has travelled to Moominvalley in hope of meeting the Moomin family, only to discover they have left—a symbol of Tove's inability to connect with the 'happy feeling' that inspired her to write their stories by that stage (2012). Due to the enormous popularity of Tove's creations, many fans have travelled to Finland to connect with the landscape in which they were created, and a significant tourist industry has developed around them, notably *Moominworld* (https://www.moominworld.fi/).

Other wainscot societies tend to be less benign and family-friendly, suggesting the marginalised and disenfranchised, and reifying the very real sections of society who are forced to live in slums, shanty towns, and ghettos. We see this more transparently in the homeless community of New York depicted in Terry Gilliam's Urban Fantasies, *The Fisher King* (1991), and *The Imaginarium of Doctor Parnassus* (2009). In Neil Gaiman's *Neverwhere* (novel and original TV series, 1996; podcast, 2013), a supernatural society is revealed to live beneath London in a literalised version of underground tube stations. In Clive Barker's *Nightbreed* (1990), the protagonist, a young man, stumbles upon a monstrous wainscot society dwelling in a place called Midian, 'where the monsters live'. This has now become a familiar trope, depicted in *Doctor Who* (2005–), *Carnival Row* (2019–2023), and in the questionable representation of otherness in *Bright* (2017)—an 'odd couple' cop buddy movie where orcs are crudely characterised versions of BIPOC communities. An ostensibly more diverse wainscot society (at least tokenistically) can be found in the Harry Potter series (1997–2007), JK Rowling's world of wizards, who live side-by-side (and sometimes with) 'muggles', or non-magical humans. The wizard academy, Hogwarts, is accessed via a special train caught from Platform 9 ¾ at King's Cross station in London, and Diagon Alley—a street where magical goods can be purchased—is hidden to Muggles and accessed behind the Leaky Cauldron pub, 'somewhere near Charing Cross Road'.

Beyond the cosy comforts of YA Fantasy (Harry Potter belongs more to the tradition of Boarding School Narratives, such as *Tom Brown's Schooldays*, *Just William*, *St Trinian's*, and *Malory Towers*, than other occult bildungsroman such as *The Dark is Rising Sequence* by Susan Cooper (1965–1977), *Summer Magic/The Journal of Luke Kirby* (1988–1995), or Neil Gaiman's *The Books of Magic* (1990–2000), exists Anne Rice's darker 'Immortal' universe, which began with *An Interview with the Vampire* (1976), and was later dovetailed with *The Mayfair Witches,* which started with *The Witching Hour* (1990), by AMC in 2022, hoping for a successful extended universe franchise. Much emulated since, especially since the success of so-called Supernatural Romance, or 'Romantasy', many of the tropes are satirised by Jermaine Clement and Taika Waititi in *What We Do in the Shadows* (2014), which spawned a popular TV series (2019–2024). Ben Aaronovitch's *Rivers of London* series (2011–) is a mash-up of the tropes of crime fiction and Urban Fantasy, featuring Police Constable and Apprentice Wizard, Peter Grant. Jim Butcher's *The Dresden Files* (2000–) ploughs a similar furrow, albeit with a grittier, American noir spin. Jasper Fforde's whimsical Thursday Next series (starting with *The Eyre Affair,* 2001) is a metanarrative largely set in the pocket universe of BookWorld, where literary classics take on a reality and the story mechanics of narrative generation are reified. It is written in a style that largely seems a pastiche or tribute to Douglas Adams' *Dirk Gently's Holistic Detective Agency* (1987).

Wainscot narratives rarely seem consciously environmental, except when they overlap with polders and become inherently vulnerable, and thus metaphoric of our own precarious position. The cross-fertilisation between paradigms becomes even more nuanced in Crosshatch narratives, or palimpsests, and it is to these we turn next.

Palimpsests

There are some Urban Fantasies where the borders between the magical and the mundane are more subtle—not delineated by geography, but sometimes by a character's state of mind. In these stories the Fantastic is another layer of reality that can be accessed at any point. Examples of this kind of Clutian 'crosshatch', or, to use my term, Palimpsest Fantasy, include *Lanark* (1981) by Alasdair Gray, which is ostensibly set in a fantastical version of Glasgow, called Unthank. Subtitled 'A Life in Four Books', Gray shuffles these into a nonlinear order—Book Three, One, Two, Four—as part of his strategy of cognitive estrangement, so that we begin in Unthank before discovering it is a 'shadow' version of Glasgow. Here, there is no sunlight, and citizens grow strange protuberances. The protagonist in these sections is called Duncan Thaw—a disaffected malcontent unable to connect with his peers. He suffers from a skin disease which results in 'dragonhide'. Thaw is eventually devoured by this shadow city, excreted into its bowels, awakening in a nefarious Institute, where hopeless cases are used as food and power. Books One and Two comprise a more naturalistic rite of passage into manhood, clearly drawing upon Gray's art college days, and yet by subtle increments—what could be called an 'endarkenment'—we discover the two cities connect in mysterious ways. In Book Four, Thaw manages to escape the Institute only to discover Unthank on the brink of collapse. After a 'heart of darkness' journey in which Thaw meets his creator (the author), he realises he can do nothing to prevent this fate. From a hilltop cemetery, he stoically witnesses the city destroyed in an apocalyptic fire and flood. Gray's singular novel is inflected with many of the doomist concerns of the late 70s/early 80s zeitgeist (also seen in *Apocalypse Now*; *Doomwatch*; *Survivors*; *Edge of Darkness*; *Quatermass: The Final Sequel*; *The Day After*; *Threads*; *The Cursed Earth*, and *Apocalypse War* storylines in 2000 AD's *Judge Dredd*, and others). It is not environmental in its concerns, other than depicting the city as an organic, semi-sentient entity, and yet that alone is enough to destabilise perceived reality and make us apprehend urban spaces in new ways. Thematically, *Lanark* could be said to explore ideas around 'nature versus nurture', and how we are shaped by the environments in which we live.

The oeuvre of the late David Lynch (1946–2025) explored the Urban Weird in much of his considerable output of film, TV drama, art, and music, but perhaps most influentially in *Eraserhead* (1977), *The Elephant*

Man (1980), *Blue Velvet* (1986), *Twin Peaks* (1990–1991; 2017), and *Mulholland Drive* (2001). The film critic, Peter Bradshaw, in his tribute to Lynch shortly after his death, summed up his distinctive brand of Fantasy and his subversion of the American dream: 'Lynch was a film-maker who found portals to alternative existences and truffled in them like they were erogenous zones, moist orifices of existential possibility. He was the great American surrealist, but his vision was so distinctive that he became something other than that: a great fabulist, a great anti-narrative dissenter, his storylines splitting and swirling in non sequiturs and Escher loops' (2025). Lynch's camera delved into the smeuses of our collective subconscious, and found a fascinating and terrifying allure in the architecture of the Id.

The City and The City (2009) is a Weird Fiction novel by China Miéville, which sets a police procedural in a world where two versions of a city (Besźel/Ul Qoma) co-exist and overlap. To help linguistically frame this Palimpsest, the author creates his own glossary, including Alter ('adjective which designates the other city: Ul Qoma for Besźel), and *Brutopically* ('adverb that designates the physical proximity of a place located in Besźel and another located in Ul Qoma'). A Marxist scholar, Miéville is interested in notions of erasure and cultural myopia—how we filter out what we do not want to see: be it the homeless, ethnic minorities, or crime. For him, the heterotopias of the city are places of imaginative possibility that challenge xenophobia and jingoism. His novels depict a fecund cultural biodiversity: Catholic spaces of ontological generosity.

A YA Fantasy novel that does this also is *Tyger* (2022) by SF Said. This is the fourth collaboration between SF Said and the multi-talented artist Dave McKean (preceded by *Varjak Paw*, *The Outlaw Varjak Paw*, and *Phoenix*), and the integration of text and illustration is an important element of the reading experience, one that enlarges the imaginative possibilities by breaking up the textual plane, and rewilding the page. Although ostensibly in the 'YA' category, *Tyger*, like Philip Pullman's *His Dark Materials* and *The Book of Dust*, has a genuine crossover appeal that would connect with adult audiences as well as with younger readers. In a similar way to how Pullman draws upon John Milton in his series, Said here draws heavily upon William Blake—not only in the titular 'Tyger' of the title (an anthropomorphic and archetypal presence that looms large in the story and in the textual plane—akin to how the armoured polar bear, Iorek Byrnison, does in Pullman's 'dark materials' universe), but in other intertextual allusions to the Lambeth-based artist and poet. The main antagonist is Urizen, Blake's god of reason—and the Tyger is almost an embodiment of Los: his blazing deity of the imagination. Yet beneath this Manichaean conflict there are several allusions to Blake's Songs of Innocence and Experience: Said draws upon the poet's iconography (the lamb; the chimney-sweep; shepherds) and cosmology. For *Tyger* is set in a parallel contemporary London—one of many in Said's quantum multiverse—in which the British Empire still exists, slavery has not been abolished, and the sprawl of the city has been checked by an authoritarian, xenophobic regime. In this other London, a secret gateway exists, connecting all the worlds. In rendering this, Said also draws upon the symbolism of another of Blake's poems, 'Jerusalem':

> I give you the end of a golden string;
> Only wind it into a ball,
> It will lead you in at Heaven's gate,
> Built in Jerusalem's wall.

This vast cosmic drama is focalised most effectively through the viewpoint character of Adam Alhambra, a young, frustrated artist who stumbles upon the Tyger in a ruin while delivering packages for his parents' tailoring business. *Tyger* revels in the 'edgelands' of urban spaces, and foregrounds the marginalised: for both Adam and the spirited young writer, Zadie, he meets are from Muslim families. The real boldness and strength of Said's novel is the way it makes such alterity not tokenistic, but central to the characters and to the plot. *Tyger* fearlessly addresses the topical issues of racism, inequality, and divisive manipulation of the masses. As such, there is both an ageless classic quality to the book and a painfully resonant one. There is also an ecological subtext to the story—'tygers' being thought extinct in the storyworld (and this world's most endangered big cat), along with many other animals: a few token species are kept in the chief villain's zoo—a symbol of all those trapped by the 'mind-forg'd manacles of man'. At one point, the Tyger seems to address the reader: 'What kind of world could kill them all? And what kind of future can such a world have?'

The City We Became (2020) by N.K. Jemisin, developed from her short story, 'The City Born Great' (first published in her 2018 anthology, *How Long 'Til Black Future Month?*) imagines how cities become sentient through human avatars. When the avatar of New York City falls into a supernatural coma and vanishes, the avatars of the Five Boroughs of Manhattan, Brooklyn, The Bronx, Staten Island, and Jersey City unite to fight their common enemy. Jemisin's novel, written throughout the first Trump Administration (2016–2020), can be seen as a cri-de-coeur of united defiance against a toxic, xenophobic regime, showing how Fantasy *can* be politically engaged. The success of the novel shows how its topical resonance and empowering message struck home with many readers. Again, as with Miéville and Said, Jemisin argues for the positive largesse of urban life, for not just tolerance, but social justice and equity—a redressing of the many imbalances of privilege and hegemonic power.

The Sunken Land Begins to Rise Again (2020) by M John Harrison is a double portrait of Brexit Britain and the isolated lives within it. It explores epistemological uncertainty: the subsidence of identity and slippage of meaning in the everyday. The author refuses the comforts of closure. He jabs a stick at the disturbing folklore of the urban and the rural: the little myths and rituals we furnish our lives with. With surreal disquietude the population seems to be devolving into some kind of aquatic throwback—a deliquescent infantilism: Brexitanian water babies.

Harrison is the master of the memorable sentence—with exhaustive invention he continually fashions glittering shards of prose, like freshly unearthed microliths glinting in the sun. And yet the prose is rarely hard to follow—the obscurity is in the precise nature of things, which he continually destabilises. He offers a sharp-eyed portrait of contemporary life in a small, damp island off the Atlantic seaboard—isolated by its own cultural solipsism, living off former glories that long ago lost any kind of global cache beyond the tourist cash cow of 'heritage'. But unlike the regulated, signposted trails that the characters come across in their almost somnambulistic wanderings, there is minimal signposting here. The reader, like the two main protagonists, is left to flounder in their own strange, twilight life. Shaw, unstable and undermined by the unreliable narrations of his uncaring care-home bound mother, lives in London and ends up working for a purveyor of a curious aquatic conspiracy; his sometime lover, Victoria, relocates to the 'sticks', inheriting her mother's house in the Welsh Marches—an ostensibly idyllic rural 'escape' that comes with its own set of problems. The anti-pastoral disenchantment is mirrored by the enchantment of the urban. Domestic and public spaces are destabilised by increasingly weird happenings. Nothing is quite what it seems, and it is almost impossible to grasp entirely what is going on—like the protagonists, we are left in the dark, refused admittance to the occult inner circles, and continually thrown by the disquieting tides around them. Harrison's agenda seems to be to induce ontological and epistemological anxiety: can we trust anything, even ourselves? Nothing and nobody is quite what it seems. Something chthonic emerges in the interstices of people's atomised lives. The characters drift apart, missing each other's calls or letters, or choosing to ignore them. And in the gaps the uncanny enters in—creating the unheimlich within the home, until what 'should' be native becomes alien, and vice versa. We become othered from our own lives.

Finally, in our foray into Urban Fantasy, let us turn to scenarios in which nature has fully reclaimed the city.

Nature Victorious

In some Urban Fantasies nature is no longer marginal, only found in the edgelands, or in metaphors of diversity—it has fully reclaimed the city. The 'rewilded city' has become a trope in recent years thanks to the proliferation of post-apocalyptic narratives, especially in serialised dramas, film, and computer games. In *Planet of the Apes* (the original film series, TV series, and film remakes and extensions of the franchise), a group of astronauts crashland on a planet ruled by highly evolved simians, only to discover that it is in fact a Future Primitive Earth, decimated by a nuclear holocaust. In *I Am Legend*, Richard Matheson's 1954 post-apocalypse novel, adapted as *The Last Man on Earth* (1964), *The Omega Man* (1971), and *I Am Legend* (2007), the solitary protagonist, the apparent sole survivor of a pandemic that has turned humans into 'vampires' (although they share similar traits to zombies, and became an influential ur-text in the zombie apocalypse subgenre) wanders eerie abandoned cities accompanied by his beloved dog. Matheson's novel is itself a variant on the 'Last Man' trope first pioneered by Mary Shelley in her titular novel of 1826, itself an expansion of Daniel Defoe's *Robinson Crusoe* (1719), a literary rendition of the 'marooned' narratives of naval oral tradition that began to proliferate in the Age of Exploration. While the 'last man' trope went on to influence John Wyndham's *The Day of the Triffids* (1951), and Boyle/Garland's *28 Days Later* (2002)—at least in its opening sequence of a patient awakening to an eerily quiet and abandoned London, the survivor and his canine companion became a trope of itself, reified in Harlan Ellison's cycle of stories, *A Boy and His Dog*—first published as a short story in 1969, then continued in other stories, and a comic book series, and adapted into a film in 1975. Set in a post-apocalyptic North America decimated by two more world wars, it follows a 15-year-old boy called Vic and his telepathic dog, Blood, as they set off across the wastelands in search of sustenance and companionship. In Ellison's outdated, somewhat misogynistic apocalypse, there is a scarcity of women, because most were wiped out when US cities were bombed while their menfolk went off to fight in the escalating conflicts. The adolescent protagonist is driven blindly by Eros and an overwhelming desire for sex, while avoiding the Thanatos of roverpacks of scavengers and mutants. In a transference of attributes, Vic is more adept at foraging for food, while his literate, misanthropic canine companion aids his human by helping him find a female. Although its crude treatment of gender would not be acceptable today, trace elements of Ellison's story can be seen in George Miller's *Mad Max* (1979) and the *Fallout* (1997–) computer game and screen adaptation.

A more subtle and complex variation of the 'boy and his dog' trope can be found in Russell Hoban's *Riddley Walker* (1980). Written in a phonetic form of pidgin English created by Hoban to suggest a future primitive breakdown of society and language, approximately two thousand years after a nuclear apocalypse, it tells the story of the 12-year-old protagonist, and how he stumbles upon a plot to recreate a weapon of the 'Bad Times'. As with *A Canticle for Leibowitz* (1959) by Walter M. Miller Jnr, a Church and State have combined into a geriatric Theocracy full of strange taboos and a bastardised version of the past based on misinterpreted fragments. The trickle of knowledge preserved from nuclear apocalypse is controlled by an anti-Enlightenment elite. In Hoban's future England 'Inland', sentient dogs become the infrequent allies of Walker as he traverses the nightmarish wastelands in imagery reminiscent of Tarkovsky's *Stalker*.

In an early *2000AD* story, *Disaster 1990!* (Gerry Finley-Day's prequel to *Invasion!*), shotgun-wielding lorry driver Bill Savage survives in a flooded London, travelling around in his DUKW, echoing the aesthetic of Richard Jefferies' *After London* (1885). The Tom Hanks vehicle *Finch* (Sapochnik, 2021) imagines a similarly bleak scenario in a world made hostile by extreme heat—in this case, the lone survivor Finch travels around in his customised RV with his self-made AI-sentient android for company (made to protect his beloved dog).

More positive visions of the future (in terms of biodiversity at least) can be found, perhaps surprisingly, in some of the novels of J.G. Ballard. Ballard used his own childhood experiences in the decaying splendour of the Post-colonial Far East to shape his dystopian vision of the future in his quartet of environmental disaster novels: *The Wind from Nowhere* (1962); *The Drowned World* (1962); *The Drought* (1964); *The Crystal World* (1966). His later novels explored a similar aesthetic of entropy and ennui. *The Day of Creation* (1987) explores similar territory, and yet the feverish subtropical aesthetic in which the vestiges of civilisation are overwhelmed by the ever-encroaching jungle—clearly drawn from his childhood memories of the Far East—is not entirely a bad thing, certainly not from a Deep Ecology non-anthropocentric perspective. Nature is healing and taking back what once was Her's. Such narratives eerily foreshadow the scenes of empty cities being recolonised by nature during the Covid-19 lockdowns, and also offer strangely reassuring (if misanthropic) vistas in an overpopulated world. If we agree with Sartre that 'L'enfer, c'est les autres' ('Hell is other people'), then a depopulated post-apocalyptic future offers a perverse kind of consoling fiction (to some at least).

George Stewart's 1949 novel, *Earth Abides*, is singular in both senses—it is the only SF the University of California Professor of English ever wrote, and also a remarkably prescient and deeply moving epic. Set in the aftermath of a virus that decimates the global population—the Great Disaster that derails the human project catastrophically (at least in terms of what we think of as 'civilisation')—the opening chapters depict an eerily quiet and depopulated land that could easily be one in lockdown. Yet as the protagonist, Isherwood Williams (or 'Ish' as he becomes known) makes his solitary way back from the wilderness where he had been undertaking field research, it soon becomes apparent that a devastating plague has swept the land, leaving fly-ridden corpses in lonely gas stations and mummified ones in the desert, and rendering the former population clusters of cities as no-go zones. And the near mass extinction event of humankind allows for a rewilding of America, in a similar way to how Victorian nature-writer Richard Jefferies imagined a 'wild England' in his post-apocalyptic novel of 1885, *After London*. Yet, unlike in Jefferies, where the first half of the novel is a detailed natural history survey sans character or plot, in Stewart's narrative, Ish is our viewpoint character who has agency. We experience this biological apocalypse through his thoughts and senses—an academic, he reflects upon what he beholds stoically. Used to his own company and absorbed by his own preoccupations, he is able to cope with a depeopled California, until finally jarred out of his solipsism by first a dog, and then by chance encounters with the diseased, deranged, or decadent few who have also survived. He embarks upon a bleak road trip to the East Coast, only to be unimpressed by the remnants he encounters. Returning to the West Coast and his former childhood home, he settles down to a quiet life, until circumstances dramatically change. What is refreshing about Stewart's post-apocalypse is the anthropological approach he takes in charting the vicissitudes of the remaining survivors. He takes the long view of history, and prophesies a circularity to it—the survivors subsist upon what they can scavenge, but eventually the shelves empty or are overrun by the swarms of ants, rats, and feral canines, and the scattered tribes regress into a future primitive state. The novel shows its age in some places—most notably in its problematic descriptions of people of colour, the handicapped, and women. And yet Stewart nearly redeems himself by lauding the main female (and mixed race) character—who is shown to have greater strength and stamina than the men. She is rather put on a pedestal and is frequently referred to as the 'mother of nations'—and so this idealised feminine is just as problematic in its own way. Stewart also is far off the mark in his disavowal of climate studies as being of any relevance to future life on Earth: 'Climatic change was not a practical problem.' Yet for a novel written in the late 1940s, we can hardly blame the author for *that* blindspot, and in many ways, Stewart's sole foray into the speculative is a seminal work of Climate Fiction, and in that sense, it is far ahead of the curve. It won the first International Fantasy Award in 1951. So, despite its weaknesses of representation, the novel has many strengths—not just the breadth of its vision, but in its non-anthropocentric shifts, and its proto-ecological tone. It foregrounds the importance of environment, and exhorts (of the earth), 'There is nothing else by which men live'. Stewart emphasises the Earth will survive us, and is indifferent to our plight. He destabilises our imagined position as the pinnacle of creation; he also challenges the vanity of ambition, the empty intellectualism of academe (whenever it ceases to have practical purpose), and the myth of progress. All that matters, he seems to infer, is our immediate community of connections, the family (or 'Tribe' in its extended form), our inner resilience, adaptability, and capability. Simple skills of survival become more important than the vainglorious dreams of betterment and posterity. And yet although this heartbreakingly charts the end of the Enlightenment Project and Western civilisation's brief moment in the sun, this is ultimately a humanist and humanitarian novel, and there is deep poetry and compassion here—in the poetic, pseudo-Biblical epigraphs; and in the loving record of marriage and friendship. A haunting vision of a plague-stricken America, there is nevertheless a quiet beauty here that lingers long after the book has been put down.

A diversification of post-apocalyptic scenarios was much-needed after decades of scenarios in which solitary white males are the protagonists. This started to happen from the late 20th century, but took off in the early 21st century as publishers of Fantastika deliberately sought to diversify their lists and improve representation in their novels. We shall look at this in more detail in the next chapter.

5 21st-Century Fantasy—Fantastic Ecologies in the Capitalocene

This chapter argues that Fantasy of the 21st century cannot be written without some acknowledgement of the multiple challenges we face—what has been called the 'age of polycrisis' by Adam Tooze (2022), and the 'metacrisis' by Manda Scott (2023)—geopolitical tensions; questions of representation and social justice; religious and racial intolerance; the Climate Emergency. Here we will focus on one key aspect that has become increasingly tangible in recent works in the field: how modern Fantasy is engaging with Climate Change (and all the attendant problems of anthropogenic environmental crisis). In works such as the *Broken Earth* trilogy: *The Fifth Season* (2015), *The Obelisk Gate* (2016), *The Stone Sky* (2017); *Lagoon* (2014); *Zoo City* (2010); *Who Fears Death* (2010); *Cloud Cuckoo Land* (2021); *The Bone Ships* (2019); *Cage of Souls* (2019); *Babel* (2022), and others we see a new awareness, sensitivity, and robustness in addressing these challenges. Yet alongside such texts we also see the rise in transmedia storytelling and 'Convergence Culture' (Jenkins) with billion-dollar Fantasy franchises (*The Lord of the Rings*; *Star Wars*; *Harry Potter*; *MCU*; *Game of Thrones*) creating 'universes' and whole creative 'ecologies'. While capitalising upon a natural aesthetic, do such 'apex-narratives' risk becoming a part of the problem that is destroying the very wonders they celebrate via their huge carbon footprint? Can these worlds—of Fantasy and the Real—be reconciled, or are they on a collision course to mutual destruction?

21st-Century Fantasy—Defining Traits

So, what, if anything, defines 21st-century Fantasy? In the first quarter of the new century much has been published (including countless self-published, and now AI-generated books), but in work that has stood out, in terms of critical reception and literary prizes, some traits have started to emerge. Firstly, nothing written since the cataclysmic events of 9/11 can fail to reflect, in some way, that generation-defining trauma and paradigm shift.

DOI: 10.4324/9781003463665-6

Of course, this might have resulted in hawkish militarism, jingoism, and Islamophobia in the worst manifestations of the post 9/11 zeitgeist, but eventually we started to see in the first two decades a healthy shift away from the old hegemonies of the 20th century—the Eurocentric, medievalist, white, cis-gendered, heterosexual, and colonial—with greater diversification, in terms of both an active (if somewhat limited and tokenistic) promotion of BIPOC authors by publishers, and better representation within the storyworlds created. There is a greater willingness to deconstruct and queer the tropes of Fantasy—with a pushback informed by the campaigns of Black Lives Matter, #MeToo, and the LGBTQI+ Communities. Alongside this, we start to see in 21st-century Fantasy a greater awareness of environmental issues, which of course is our chief concern here—but there is also a growing consciousness that these issues are inter-related and 'intersectional', to use Crenshaw's term (2017).

Now let us consider some key themes and examples of 21st-century Fantasy. To begin with, we shall examine a trend that ostensibly seems regressive, but actually can get to the core of humankind's historical problematic relations and contemporary disconnect with nature.

Blood Harvests: Eco-Paganism in Fantasy

Ever since Robin Hardy's iconic, influential Folk Horror film, *The Wicker Man* (1973), the trope of an isolated, failing community that requires some kind of blood sacrifice to thrive again has developed in TV and radio dramas, films, fiction, and graphic novels. It has become so familiar to audiences that it has become ripe for satire, for example, *Get Away* (dir. Haars, 2024). One could pathologise the root of this trope back to James George Frazer's monumental, if flawed, study of comparative religions, *The Golden Bough* (1890), which sought to identify a monomyth—a vegetative cult of death-and-resurrection—from the world's myths, legends, and folklore. One could even argue that the trope of Pagan 'human sacrifice' can be traced back to Roman propaganda, primarily *The Conquest of Gaul* (trans. 1982) by Julius Caesar, which depicted the Keltoi and the Picts as woaded, blood-thirsty 'savages' who needed conquering and civilising—the same narrative of genocidal nations intent on extending their territories (and acquiring precious resources) ever since. Yet it was Hardy's film that brought the idea into popular modern consciousness. As the idea (or rather misconception, as seen in the fallacious 'Slaughter Stone' of Stonehenge, which Thomas Hardy used in the climax of *Tess of the D'Urbervilles* (1891), leaning into the Victorian fallacy that it was where the Druids—the Iron Age priest caste who were once mistakenly associated with the Neolithic monument—conducted their sacrifices due to the red crust upon it, which turns out to be naturally-occurring lichen) has dramatic potential, if very little historical basis, it was seized upon by writers throughout the later decades of the 20th century in dramas such as *Robin Redbreast* (Play for Today, 1970), where a heartbroken woman called Norah is unwittingly lured to a remote village to breed a 'new Robin' for a disturbing Pagan cult; John Boorman's *Excalibur* (1981), which depicted a Wasteland scenario, combining elements of the Parsifal legend and the Frazerian idea that the 'King and the Land are one'; and Richard Carpenter's *Robin of Sherwood* (1984–1986), which placed the titular character, played by both Michael Praed and Jason Connery as two versions of the folk hero, as 'Herne's Son', Herne the Hunter being his shamanic mentor figure and ancient guardian of the land. Robin's death and resurrection at the end of Season Two and start of Season Three capitalised upon Frazer's idea, expanding Robin Hood from historical bandit, to the mythical figure of the Green Man, an archetype which took off in the late 80s and 90s, via his adoption by the so-called New Man movement pioneered by the likes of Robert Bly, and by the New Age, via the works of William Anderson (1990) and John Matthews (1993), the latter collaborating with Mark Ryan, an actor and fight co-ordinator from *Robin of Sherwood,* on the popular *The Wild Wood Tarot* (2011).

All of this fed into the manifestation of Eco-paganism in Fantasy in the early 21st century. It walked onto the big screen in Peter Jackson's *The Lord of the Rings* trilogy (2001–2003), notably in his depiction of 'The Last March of the Ents' in *The Two Towers* (2002), where Treebeard, when beholding the destruction Saruman had inflicted on the forest abutting Isengard, rumbled his threnodic cry (in the Boyens and Walsh screenplay): 'Many of these trees were my friends, creatures I had known from nut and acorn; many had voices of their own that are lost forever now.' To which Tolkien, whose love of trees stemmed from his childhood mythscape, adds, 'And there are wastes of stump and bramble where there were singing groves.' Well aware of the Brythonic tradition of magical literature as depicted in *Y Mabinogi* (probably reading Lady Charlotte Guest's late 19th-century version) and the 14th-century *Hans Taliesin,* Tolkien would have been aware of the poem, 'Cad Goddeu', the 'Battle of the Trees', in the latter, and was probably drawing upon this, as well as the famous scene from Shakespeare's *Macbeth* when the Thane of Cawdor is terrified that the Witches' prophecy of his doom was coming true: 'Macbeth shall never vanquished be until/Great Birnam Wood to high Dunsinane Hill/Shall come against him' (IV: 1.105–107). Believing he was safe 'Till Birnam wood remove to Dunsinane' (V: III), Macbeth then sees it before his very eyes, albeit via the use of foliate camouflage by the approaching army of his blood-enemy, Macduff, who hold branches before them. This sign he sees as an omen of his inescapable doom—the karmic comeuppance for his overweening ambition and murder of his king, Duncan. Tolkien and

Jackson's representation of non-anthropocentric ecological justice is rarely seen elsewhere (Alan Moore's reimaging of DC's *Swamp Thing*—running 1984–1987—being a rare exception). The Caesarian/Frazerian trope of human blood sacrifice is far more common, encountered numerous times in the slew of Folk Horror that gained popularity in the second decade of the new century, in films such as *The Ritual* (2017), *Apostle* (2018), *Midsommar* (2019), and the three-part series, *The Third Day* (2020).

More positive iterations of Eco-Paganism are rarer—Folk Horror typically portrays the countryside as a place of discomfort and danger (for unwitting 'outsiders' usually from towns and cities), where ancient grudges are played out by isolated, regressive communities. This anti-pastoralism is a very urban perspective, and one that portrays Paganism in a negative light.

However, there is an exception in Mackenzie Crook's *Worzel Gummidge* (2019-), which celebrates the benign, healing aspects of the natural world through its soothing imagery of the English landscape, and cast of bucolic characters. Crook's update of Barbara Euphan Todd's series acknowledges the impact of anthropogenic Climate Change, although never explicitly, through the disrupted cycle of seasons noticed by the titular scarecrow and his magical fellowship of feather (the crows) and straw (scarecrows)—alongside the farmer and his wife—indeed it becomes a central plot thread. It is down to the scarecrow community to set matters right through some enchanting old magic, in which they sing and dance the health of the land back again, in Ten Acre Field at least.

A darker iteration of this symbiotic relationship between an archetypal vegetative figure and the health of the land can be seen in *Inside No 9*—'Mr King' (S7: E2), in which an irascible schoolteacher played by Reece Shearsmith, moving to a Primary School in rural Wales, finds himself becoming the folkloric sacrifice, in the tradition of Sergeant Howie, Edward Woodward's character from *The Wicker Man*—this time, inflicted upon him by the children.

A more nuanced use of the tropes of Folk Horror can be found in *Thistlebone*—the *2000 AD* comic series created by T.C. Eglinton and Simon Davis (serialised progs 2135–2144, 2019, 2021; collected in graphic novel form 2023). Here, Avril Eason, the escapee of a cult in the village of Harrowvale, who returns to the site two decades later to come to terms with what happened, eschews the usual sacrificial victim role, even though she has clear psychological damage due to what she experienced (the PTSD depicted provides a refreshing note of realism, while also making Eason an unreliable narrator). The devotees of an ancient Pagan deity, who resides with the woods that surround Harrowvale, chant the following:

> Beneath the waxing moon, ripen kin and swell the fruit,/Hang heavy the branches and drink deep the root/Let fox and stag lead the chase, to

> bless these fields and bless these folk!/For what we've taken we now give back, to Old Thistlebone of the Oak!
>
> (2023)

The cult suggests a deep connection to the land and its cycles—a neo-animist rustication that the urban interlopers find deeply threatening. Not only is there 'nowt so queer as folk', but also the ways of the countryside are opaque, trapped in a timewarp of history and tradition, and mercilessly bloody to outsiders.

'Pilgrim' is a BBC radio drama series by Sebastian Baczkiewicz about an immortal protagonist, William Palmer, who, in the tradition of the 'Wandering Jew', and *Melmoth the Wanderer*, is cursed by the Grey Folk to wander the Earth after denying the existence of the Otherworld on the road to Canterbury in the 12th century. In 'Belle Meadow Fayre' (2024), the Ecological Imaginary is foregrounded in a plot involving the abduction of a John Barleycorn figure, Old Johnny John John, by a rogue member of the 'Grey Folk', in his adopted home village of Woodwarthing. Without the abductee, the annual Belle Meadow Fayre—a ritual to mark the end of Autumn—cannot take place, and harmonic seasonal synchronisation cannot occur.

In Fantasy fiction, we see some of these eco-Pagan tropes being played out in novels by Robert Holdstock, Naomi Novik, and Zoe Gilbert. In *Avilion* (2009), Robert Holdstock's last novel, the author returns and concludes his Mythago Wood Cycle, and brings his unique syncretist spin to his sylvan legendarium: a place of mythopoeic generation (Manwaring, 2018). After 25 years Holdstock finally picked up the storyline directly from the famously open ending of the first, which saw Steven's glimmering girl, Celtic princess Guenette, being abducted by his father (in the primal figure of the Curcuma) and taken back through the wall of fire that guards the way to Lavondyss. A coda relates how a giant came to a tall stone marking the edge of his known world, the head of a valley where he met a hunter waiting for 'the girl who came back through the fire', and this bequeaths the valley with its later name, *Iman uklyss*.

And finally, she did: Steven and Guyenne are reunited. They settle down in the valley in an old Roman villa (a mythago conjured from Steven's memory) and have a couple of children, Jack and Isobel. Avilion resumes the story at the point when this idyll is shattered. Guenette has disappeared, and Ysabel has gone to find her—heading inwards to Avilion/Lavondyss. Jack (half-human, half-mythago like his sister—both 'red' and 'green') journeys outwards to the edge of Ryhope—desperate to see the world of his father, the Lodge where it all started, the village of Cedarhurst, and hopefully find some clues that will help in the search. By the time Jack reaches 'the fields we know', it is the 'present day', possibly late 20th century (the original disappearance of the three Huxley men—George, Steven, and Christian—occurred in the late Forties). As all travellers of Faerie discover, time runs differently in each world. In an intertextual touch, Steven has quested Jack with finding his old copy of H.G. Wells' *The Time Machine* (1895) in the Lodge. Jack's presence there dislodges the ghost/mythago of his grandfather, George Huxley—and we have some insight into the original chain of events. Echoes and reflections occur a lot throughout the book—things overlap and repeat. The polder of Ryhope Wood has a porous boundary—the real world leaks into it and vice versa (one of Holdstock's extended cast from the Mythago Cycle is Caylen Reeve, a charismatic shaman figure living in the village, epitomises this). This symbiotic nature is not surprising, for Mythago Wood is a cypher for the imagination, a zone of subconscious creativity—'Avilion is what we make it' (2009, various)—and it is at its most fertile where the two worlds (waking and dreams/conscious and subconscious) meet. Holdstock is not afraid to plunder (or cross-reference) his own treasure hoard. Christian's army of the lost, Legion, marches into the book from the pages of *Gates of Horn, Gates of Ivory* (1997) and that wily Greek sea captain, Odysseus, pops up as a love interest to the young Isobel—a kind of leftover 'mythago' from the *Merlin Codex* (Holdstock's Celtic/Ancient Greece crosshatch trilogy). This self-referencing could become a law of diminishing returns, but it satisfyingly capitalises on previous books in the sequence and draws the threads together. The multilinear narrative structure is bold—intercutting between mainly Jack and Yssobel's journey, but sometimes switching to Guiwenneth's, Steven's, and Christian's. In his weaker work, this oscillation of viewpoints can leave the reader sometimes floundering. Here, Holdstock carries it off. The prose is well-honed and lucid. The characters of Jack and Yssobel are convincing—indeed, Yssobel's voice comes across the clearest, and this is more her story than anyone's. The portrayal of the older Steven and Guiwenneth is a believable depiction of marriage—albeit a mythopoeic one—and Steven's transformation into his father is touching (it could even be a gently self-deprecating self-portrait). Holdstock includes a selection of his poetry—one of which, 'Fields of Tartan', is based upon his grandfather's memories of the First World War—he incorporates this into the narrative effectively, but the personal reference gives the whole thing added poignancy. Holdstock also dedicates the book to the memory of his father, who died before it was finished: again, this fact gives the story an added depth, as the novelist depicts an otherworld where loved ones never die. It is a novel about homecoming—'What we remember is all the home we need'—and in *Avilion*, Holdstock has finally come home. This is a fantasy novel willing to depict mature relationships and shows Holdstock's maturity as a fantasy novelist, made all the more poignant for being his last.

As all great works, 'Avilion' conveys something fundamental about the human condition, why we, the human creature, need to dream, to explore, to discover, to take risks, even to suffer—if we are to truly grow:

> We held and held until we broke,
> But in the breaking, we held,
> and in the holding we will find Avilion.
>
> (2009)

In terms of the Ecological Imaginary, beyond the sylvan frame-setting, this is evidenced by the inclusion of the dryadic perspective: these other-than-human presences are given voice and agency. At one point, alarmed by the approach of Christian's devouring army of phantoms, Legion, a dryad utters:

> *Danger is coming. We will be broken. We will be fire-stuff. We will be cut. We will be broken by anger. We must protect our trees. I have lived in mine four hundred leaf-falls. I will die if I cannot drink the sap that rises at leaf-burst. All of us will die. Birds, be vigilant. Beak, claw and bird-screech will be needed.*
>
> (Holdstock, 2009: 231)

Jack and Yssobel are continually oscillating between the 'Red' and the 'Green' inside of them, as a poignant symbol of the more primal and higher evolutionary elements of our own anatomies and consciousnesses. As 'Haunter' Jack is able to access a deeper, fuller sensorium than the one available to him as a human:

> Within the body of the young man, Haunter was terrified for the first time in Jack's life. He became a forest shadow, darting through the earth, pressed against the looming boles of the forest, ascending to the canopy where the air was always fresh and moist. It was a time of terror. He engaged with what he could: the musty floor of leaves, the flower-filled dells, the animal sanctuaries within the groves, the old tracks that he knew so well by an instinct drawn from the very wood itself.
>
> (2009: 306–307)

Ultimately, it is this 'Green' side of mortally wounded Jack that must be offered to the Valkyrie/Morrigan psychopomps who escort the fallen to Avilion, so that he can continue to live—within the sustaining energy field of Ryhope at least. Meanwhile, his sister, Yssobel, dwells deep within Avilion, awaiting the return of her resurrected mother.

Zoe Gilbert's composite novel, *Mischief Acts* (2022), charting the transformations and travails of Herne the Hunter from the Early Middle Ages to the late 21st century, is a riotous celebration of a fascinating folkloric figure—re-imagining via an ecofeminist filter as a kind of accidental eco-warrior. In his 'origin story' as the King's favourite huntsman, he is fatally wounded in a hunting accident while saving his monarch's life. A mysterious figure called Bearsman restores him to life by strapping the slain stag's head to his head. Herne revives, but has lost all of his skill. He loses his position, and is found hanged; but whoever takes over his former role, as the King's huntsman, experiences the same. Eventually these lost souls are told to meet Herne at the oak where he was found hanging—and there Herne awaits, to claim them as recruits for his Wild Hunt. This spectral pack are seen on moonlit nights, riding wildly across the land, gathering in all the lost souls. So far, so dead. From this source story, Gilbert mutates the Herne legend through a flickerbook of English history, occasionally draping the hides of other tales on top, or casting off aspects that no longer seem so acceptable. At first Herne is a threat to the forest (here represented metonymically by the 'Great North Wood', a fragment of Anderida in South London that survives mainly only in placenames such as Norwood), riding roughshod over its flora and fauna, bothering nymphs, and disturbing the peace. This is amusingly related via the perspective of a river goddess, Effra, who decides to rein in Herne's wildness, only to end up frolicking with him in a nightly love chase. Herne is cast as the ultimate bad boy—mad, bad, and dangerous to know, yet darkly charismatic and irresistible to some. Herne is at times a trickster, a seducer, a rabble-rouser, and revolutionary. The quintessential green man (or woman), he invites us to connect with the wildness within. His feral energy challenges the status quo, shakes things up, frees up stagnant situations, catalyses radical actions and change. Yet his anarchy is rehabilitated here somewhat, as he becomes the defender of the wildwood and the liberties and rights it affords: the first eco-warrior, and the personification of the wild, its immune system rising up against incursions and abuses. Gilbert's repurposing of Herne in this way gives him a resonant, topical edge—yet he remains distant, as the author focuses on marginal characters who come into contact with him, attempt to fight his influence, or fall in love with him. We never gain an insight into Herne's own modus operandi. We never see inside his head—he remains the romanticised 'other', the demon lover, the wild twin (Shaw, 2023) who is eventually confronted and embraced in the final chapter. Gilbert seems to be saying there could be a little bit of Herne inside all of us, for we are all susceptible to 'mischief acts'. When we transgress, we channel a bit of Herne. Yet the more Gilbert changes Herne's shape to suit her own interest or agenda—turning him into a female highwayman, a trio of sapphic nymphs on a 1960s London estate, an 'absolutely fabulous' housewife knocking back the wine and leching over the gardener, or a young, female eco-warrior raised by wolves—the more distant Herne becomes, relegated to a loose connecting device for a series of engaging, but uneven short stories. As a novel, *Mischief Acts* (2022), hangs together less well, especially in the final chapters set in 2011 to the near-future. These chapters lack the energy, focus, and astonishing historical ventriloquism of the earlier ones, breaking down into a pidgin poetry that reads like a discount version of Russell Hoban's *Riddley Walker* (1980). Gilbert's wide-ranging approach has breadth, but not depth. What it gains in literary experimentation, it loses in narrative traction and character arc. Although versions of Bearman turn up in the different scenarios as the eternal antagonist, it is only the river goddess—the novel's riparian Greek Chorus—who seems to have the potential of a fully rounded character. Gilbert carries off her voice more convincingly than Herne's, who is often only allowed to repeat a single phrase, like some isolated Incel on a killing spree. And Gilbert's pastiches are not as committed and sustained as say David Mitchell's in *Cloud Atlas* (2004), which shows how a composite novel can hang together well, and be greater than the sum of its parts. The novel is bookended by two fictional 'lectures', which provide some interesting background and meta comments, but feel tacked on, and less intrinsic to the core narrative than the pseudo-academic framing of *The Handmaid's Tale* (Atwood, 1984) or *The Power* (Alderman, 2016). Certainly there is a lot to admire and enjoy in Gilbert's fecund, boisterous gallop through history. It certainly celebrates biodiversity and local distinctiveness, and is a praise-song to our precious forests. But as a novel it feels rather fragmentary—a rag, a bone, a hank of hair that fails to bring Herne himself fully to life. Like the 2007 Bob Dylanless drama-documentary, *I'm Not There* (dir. Tod Haynes), it feels like the main attraction is missing, and we are left with some (entertaining) sideshows.

In Naomi Novik's *Uprooted* (2015), the forest is depicted as a sinister, sentient place, akin to the Forest Perilous of the Arthuriad, and the fairy tale forest of the Brothers Grimm. Drawing upon Polish folklore, Novik sets her story in a Secondary World, but one clearly infused with a distinctive Eastern European ambience. Agnieszka lives in the village of Dvernik in the kingdom of Polnya, where every year a village girl must be sacrificed to the 'Dragon' to protect it from the 'Wood'. The Dragon turns out to be a malcontent wizard, and the Wood—not quite what it seems. Portrayed throughout as a malevolent force antagonist to humans, absorbing them into trees like so many Old Man Willows, at the novel's climax, we discover why 'nature' has turned against the humans of this world. Within the heart-tree, Agnieszka is shown the backstory that has led to this situation: the Wood-Queen married a human king, but when he died, his people turned on the wood people, causing them to become trees to protect themselves. After the humans failed to kill the Wood-Queen, she returned to her people only to find that they had turned themselves into trees. The Wood-Queen was distraught with grief and hatred for what humans had done to her people. She began sealing human beings into heart-trees in an effort to protect her people and preserve what was left of them. The Wood's corruption came from hatred and misery. Agnieszka escapes the tree and tells the Wood-Queen to heal her sister's tree with her own body. She helps the Wood-Queen change into a tree, something that the Wood-Queen had not been able to achieve alone. The Dragon doubts this will keep the forest in check. However, a year later Agnieszka has tamed the Walkers, who now help her in burning corrupted trees and in planting new, good trees, thus changing and healing the relationship between the human and sylvan realms.

A discernible 21st-century trend has been the rise in both Folk Horror and Folk Realism. Although the latter often explodes the supernatural, explaining it away as a psychological effect, the result of urban myth, mass hysteria, gaslighting, and fake news, it is worth considering due to its foregrounding of the environment. A prime example of this is Andrew Michael Hurley's *The Loney* (2015)—an atmospheric study in faith, truth, and the uncanny. It focuses on two brothers—one an elective mute—growing up in a Catholic corner of 1970s London. The universe of the novel is limited to their immediate family—the overbearing 'Mummer' and forbearing 'Farther'—and faith community, the claustrophobic, competitive parish, ridden with sins, small and large. When their much-loved priest dies, a new Father tries his best to shepherd this unruly, opinionated flock. The annual pilgrimage to a shrine on the Lancashire coast, to the lonely stretch of the title, vividly evoked by Hurley, is hoped to cure all ills—especially the mutism of 'Hanny', the narrator's brother. Instead, the trip opens up a whole can of worms. The author succeeds in creating a tangible sense of place, one that epitomises the 'English eerie' as discussed by Robert Macfarlane and others (2021). *The Loney* has a brooding sense of malevolence, as though the bleak section of coastline was the antagonist itself. Something chthonic and unwelcoming lingers there. The remoteness and neglect provide a zone of projection for the inchoate fears of urbanites (something that has been happening in English Literature since *Beowulf*). Whether the drear flatlands of the fens, the grim North Country of the Gawain poet, or the blasted heath of Shakespeare, Britain's few wildernesses offer a dark playground for imagined terrors. There are shades of *The Wicker Man* here—the spectre of dodgy Pagan rituals awaiting the unwary who stumble upon them (the pilgrims might as well have 'sacrifice' pinned on their backs). Hurley's aesthetic reminded me of many of Susan Hill's settings—lonely stretches of coastline, causewayed islets, chancy footing, and vindictive tides. Madness always in the wings, or in the annex. *The Loney* 's 'heart of darkness', Coldbarrow, could be in Hill's real estate portfolio. One could imagine *The Woman in Black* 's sister living there. The stifling religiosity, small-mindedness, and demonisation of difference were reminiscent of Jeanette Winterson's *Oranges are Not the Only Fruit* (1992). Shades of Graham Joyce's Midlands' *Gothic* can be found here also—although his prose has more warmth and humour to it—and a broader church than depicted here. Hurley has done nothing new under the sun here, but as a debut it is supremely confident, polished, and promising. His prose is unshowy, no-nonsense, and easy to read, but there are poetic flourishes—distinctive dialect, arresting analogies, striking imagery—which make it punch above its weight at times. The ending is open and unsettling. Hurley resists exegesis and closure. What we are to make of the strange events is left to us to gnaw over in the middle of the night, while the existential wind of nameless terrors howls outside.

A writer who has been exploring the dramatic possibilities of place for several decades is Alan Garner, whose lifetime's association with his beloved Alderley Edge has resulted in the literary equivalent of Nan Shepherd's deep mapping of the Cairngorms. Garner's Fantasy, which began with *The Weird stone of Brisingamen* in 1960 has become increasingly sophisticated over his long life (even within his Weirdstone Trilogy we see this: it starts out as Children's Fantasy, but the final part, *Boneland*, released 50 years after the original is written in his now trademark adult cryptic style). And unlike Folk Realism, it fully commits to the ontological possibilities of the Ecological Imaginary. Garner's storyworlds are multidimensional, transtemporal, and neurodivergent: one could even say, shamanic, especially in the later novels such as *Strandloper* (1996) and *Treacle Walker* (2022).

Alan Garner's 2003 novel, *Thursbitch,* was a long time in gestation, but like the slow processes of geology, folk memory, and love, the dilatory nature of its composition has produced something distinctive and enduring. *Thursbitch* is based on a true place and a true tale of discovery. Once, when fell-running as a younger man, Garner stumbled upon a stone in a Pennine track in Cheshire with this curious inscription:

> HERE JOHN TURNER WAS CAST AWAY IN A HEAVY SNOW STORM IN THE NIGHT IN OR ABOUT THE YEAR 1755—THE PRINT OF A WOMAN'S SHOE WAS FOUND BY HIS SIDE IN THE SNOW WHERE HE LAY DEAD

Garner has extrapolated ingeniously upon this enigma, bringing to life John Turner: jagger—a leader of horses—salt-trader, and local medicine man. The dark story was largely inspired by Garner's own experience of researching the area: finding hidden valleys and mysterious stones and knowing farmers. It is a detective story across time as the events leading up to Turner's death are revealed; in parallel with the rediscovery of the mystery by an ageing couple of the present, Ian and Sally, a priest-doctor and a geologist with a debilitating disease, probably Alzheimer's. The portrayal of this odd couple, intellectual, cantankerous, and growing old disgracefully, is both touching and tough; unsentimental in its depiction of the decay of the body and the mind. Sally is drawn to the mystery of Thursbitch as she approaches her inevitable end.

The characters of both centuries are portrayed almost purely through dialogue—realistic, if alienating, in its use of jargon or dialect. Ian and Sally's conversation is as gnomic as the Colloquy of the Sages. Garner has caught the Cheshire tongue accurately, if exhaustingly. Like all slang, it alienates the outsider. The reaction to ramblers by Ian and Sal, if funny at times, is similar to the insular community of Thursbitch: 'You're not from round here', in other words. It is a singular place, rendered vividly, and if nothing else makes one wonder what other mysteries are hidden amongst the villages and valleys of this land.

Garner has an eye for the mythic in the familiar, and here the Old Gods make their presence felt—both the Northern and the Mediterranean variety. Thursbitch tells the tale of the Thunderer and his wife, on a human level. It depicts in uncompromising and unapologetic frankness the practice of Paganism by a whole community—*The Wicker Man* without Edward Woodward! Here, in a Pennine Valley is a remnant of an ancient cult: Minoan or Mithraic, mixed with the Norse.

Yet this is a very adult tale, about growing old, the search for knowledge, the nature of time, the hardship of life, and the mystery of death. A tale for the winter years. The stark clipped language offers no easy answers. It's as though Garner has taken the edict of the Northern poet Basil Bunting's advice in *Briggflatts* (1966): 'Words are too light, take a chisel to write.' There is a hard poetry about the diction, like the Pennine landscape: an unyielding, austere beauty. It takes a while for the ear to attune to this strange music, yet it is worth the effort. For Thursbitch is the real thing: a glimmer of a mystery immanent in the mundane, with salt-of-the-earth people interfacing with the divine among the ruins of time.

The furrows ploughed by Garner have provided fruitful ground for later Northern writers.

David Barnett is a productive and accomplished writer in other forms (journalism; comics; rom-coms), but *Withered Hill* (2024) is his first foray into prose Fantastika—a full-bloodied Folk Horror novel set in a fictional Lancashire village run along Pagan principles of the Hammer House of Horror variety. Isolated from the modern world, but still connected to it, hard to find and even harder to get out of, the ominously named village of Withered Hill will never be found on any map, but could be triangulated by the bearings of *The Wicker Man*, *Midsommar*, and *The Prisoner* (with a dash of *The League of Gentlemen*). The village is preternaturally blessed with fecundity and rude health in all that grows and lives there, and the locals are especially fond of observing the ancient festivals and attendant rites—these two aspects of village life are symbiotically connected. Because the good harvests and bucolic idyll enjoyed by (most) of the inhabitants come at a price. Cthonic forces needed to be placated. Into this stumbles (naked, dazed, and covered in mud and leaves) Sophie, a thirty-something Gloucestershire-born Londoner on a slippery slope of alcoholism and loneliness after seeing her friends get married and move away. Less Bridget Jones, and more Amy Liptrot (who wrote about her own descent and escape to a remote community in *The Outrun* (2018), adapted into a movie with Saoirse Ronan), Sophie turns up in the village with memory loss. The novel is a split-narrative, alternating between 'inside' (the village) and 'outside' (London, etc). Barnett handles this balancing act well, shuffling the chronology around but never confusing the reader as each section is clearly identified with a place and day count. Incrementally, the backstory is drip-fed, and the mystery of how Sophie ended up in the village is revealed. Characters are convincingly drawn, the dialogue rings true, and each scene is vividly brought to life. It would be so easy for such territory to seem hackneyed and formulaic—allowing readers to play 'Folk Horror' bingo as they tick off the tropes—but Barnett gives the subgenre a fresh energy. Unlike its more respectable cousin, Folk Realism, Barnett goes full tilt into the Folk Horror, providing many visceral thrills along the way. Ancient genius loci need appeasing, but how much of her humanity is Sophie willing to forsake to escape? Gory, sexy, and good fun, *Withered Hill* feels very grounded, inflected with a vivid sense of regionality.

David Lowery's haunting, hallucinatory re-imagining of the 14th-century Middle English verse romance, 'Sir Gawain and the Green Knight', *The Green Knight* (2021) was delayed from its original May 2020 release because of the Covid-19 pandemic: the ultimate ecoGothic trope. Fended off like the fatal blow of the antagonist several times since, it has been worth the wait. Lowery has adapted the source text in a structurally bold and visually stunning way. It makes striking acknowledgement of textual sources—suggesting through a rapid flickering of fonts the many versions and variants. 'Gawain' is a cultural virus that has mutated through the centuries, being re-translated and retold in myriad forms. There have been scholarly and poetic tellings from Tolkien, and the former poet laureate, Simon Armitage; theatrical and operatic versions—most notably Birtwistle's; adaptations for the small and silver screen (most notably in David Rutkind's faithful 1991 version); and numerous usages of key elements of the story in comic book, computer game, and TTRPG. But Lowery, with his dreamlike, visionary style, has reclaimed 'Gawain' for the big screen—with a storytelling style that has its roots in European art-house cinema more than Hollywood. This feels 'wilder', more willing to take risks. It is hard to predict where it will take you, or what astonishing image will appear next. And yet there is narrative traction, and a thematic coherency about it: the leitmotif of the circle binds the film together—in the Round Table, King Arthur's crown; the sinister ritual of shadowy priestesses; Gawain's shield; and the famous 'green garter' or belt of protective spells, which is given to him by both his mother (a dominating Circe-like presence played by Sarita Choudhury) and 'The Lady'. Gawain, played with conviction and charisma by the brilliant Dev Patel, is an ignoble, compromised figure: a hedonistic, amoral Prince Hal we hope will become our Henry V. King Arthur is depicted in full Fisher King mode (by Sean Harris), and yet Gawain is no Parsifal. His sorceress mother appears to set in motion a series of events that will lead to her son's betterment, either societally or in terms of his maturation. Gawain is a pawn, but a self-aware one, at one point asking is it 'A game?' Arthur replies: 'Perhaps.' Yet the Beheading Game that is instigated by the dramatic arrival of the uncanny Green Knight at Arthur's court is deadly serious—one with inexorable consequences. A moment of valour leads to a year-long countdown to a gruelling journey into the wild north. Here, Gawain is tested by tricksters, ghosts, giants, and apparently friendly hosts, along with the more-than-human world of nature itself. Indeed, an ecolinguistic subtext rises to the surface in The Lady's extraordinary Pagan paean. After asking 'Why is he green, do you think?' in the presence of Lord Bertilak and Gawain, she discourses upon the symbolism of the colour herself:

> We deck our halls with it, dye our linens, but should it come creeping up the cobbles, we scrub it out, fast as we can. When it blooms beneath our skin, we bleed it out. And when we, together all, find that our reach has exceeded our grasp, we cut it down, we stamp it out, we spread ourselves atop it and smother it beneath our bellies. But it comes back. It does not dally; nor does it wait to plot or conspire. Pull it out by its roots one day, and the next there it is, creeping in around the edges. Whilst we're off looking for red, in comes green. Red is the colour of lust, but green is what lust leaves behind—in heart, in womb. Green is what is left when ardour fades, when passion dies, when we die, too. When you go your footprints will fill with grass; moss will cover over your tombstone. And as the sun rises, green shall spread over all, in all its shades and hues. The verdigris will overtake your swords and your coins and your battlements, and, try as you might, all you hold dear will succumb to it. Your skin, your bones. Your virtue.
>
> (2021)

As embodiment of this hue, and metonymically of all nature, it is tempting to see the Green Knight himself as the very scion of environmental justice. Yet the flawed antagonist Gawain must face feels less the vengeful face of nature, and more a moral and spiritual catalyst. In a mind-bending dilated alternative timeline, we behold a possible fate for Gawain in true 'Last Temptation of Christ' style. This is foreshadowed by the 'death' of the protagonist in the forest: the camera panning around the seasons like the rotating backdrop in the puppet show. Gawain is bound to Fortuna's wheel—a victim of circumstance until he finds his own moral compass, his inner chivalric code. After being tested by the irresistible Lady Bertilak (played with sensuous power by Alicia Vikander) and her husband (played with earthy vigour and sexual ambivalence by Joel Edgerton), Gawain finally arrives at the Green Chapel and reaches a kind of apotheosis, sitting in Buddha-like contemplation beneath the 'Bodhi tree' of the sylvan lord. What risks being an anti-climax transforms into the most astonishing sequence in which Lowery—both writer and director—strays fullest from the well-trodden woodland path. There is only one misstep in this otherwise masterful revisioning of the poem—the CGI fox, which feels like a concession to a younger audience, a stray from another kind of 'fantasy' movie (although in the original screenplay it has a far more tangible role, serving as a foil to Gawain in the wilderness). Perhaps it only jars because Lowery has otherwise served up a feast of Fantasy of the highest order, one that deftly straddles the medievalist and the modern—in music, costume, and mise-en-scène. It knowingly weaves in its sources, while simultaneously transcending them. Lowery has returned to the rootstock of his Medieval source and found new growth, making the romance of *Sir Gawain and The Green Knight* speak to the ecological and ontological concerns of the early 21st century.

* * *

Now, let us turn to one of the other major strands in 21st-century Fantasy.

As we has seen, Folk Horror can swing from the rationalisations of Folk Realism to the anti-mimesis of Fantasy. And running through all of its occult discourse of Eco-Paganism—however sympathetically it is depicted—is the blood-red stream within the green: a current that cannot be stemmed. The more it is blocked, the more it will ooze out. But what if the porousness is within an urban setting?

Heterotopias of the 21st Century: Urban Fantasy and Other Spaces

If the broad church of Folk Horror is one way in which the rural has evolved in 21st-century Fantasy, then Urban Fantasy has similarly evolved into more complex heterotopias, reflective of the complex melting pot that the majority of the world finds itself living in. Today, some 56% of the world's population—4.4 billion inhabitants—live in cities. This trend is expected to continue, with the urban population more than doubling its current size by 2050, at which point nearly 7 of 10 people will live in cities (World Bank, 2024). Cities are no longer mere backdrops where the Fantastic can irrupt and intrude—a somewhat antagonist, oppositional force at odds with the consensus reality. In these new Urban Fantasies the urban is itself a significant presence, sometimes with sentience and agency. In Philip Reeve's *Mortal Engines* (2001), whole cities are on the move, fighting for resources in a post-apocalyptic future; echoing both the aesthetic of *Howl's Moving Castle* (Wynne-Jones, 1986; Miyazaki, 2004) and John Garth, the city hunter, in Robert Holdstock's *Ancient Echoes* (1986). But where does the Ecological Imaginary reside in such narratives? Anyone who has lived in a city will testify to the presence of the more-than-human—be pigeons, urban foxes, rats, and cockroaches. The lockdowns resulting from the Covid-19 pandemic saw nature return to the silenced, emptied towns and cities in unexpected ways: goats strolling into Llandudno, wild boar running amok in Barcelona, herds of Sika deer heading downtown in Nara City, Japan, and mobs of monkeys terrorising locals in Lopburi, Thailand. Such imagery echoed countless dystopias, and can be seen in novels such as Emily St John Mandel's *Station Eleven* (2014), which offered a refreshing take on familiar territory—the weary tropes of the post-apocalyptic. St John Mandel's prescient novel depicts a severely depopulated world ravaged by a pandemic with a very high fatality rate, the 'Georgia Flu'. The main story is set two decades after 'year zero', and follows a group of classical musicians and actors around, the 'Travelling Symphony', as they take their revue from settlement to settlement. This is a powered-down world with horse-drawn carts and very little surviving tech beyond the mechanical, a return to the (re)Wilded West. Much of the country has regressed into Badlands and no-go areas. Yet while this acknowledges the usual Grimdarkian survival of the fittest, St John Mandel's novel posits a more redemptive scenario—one in which art still has a value. The central tenet of the book is summed up by a quote from *Star Trek: Voyager* ('Survival is Insufficient'), which is emblazoned on one of the wagons of the caravan—and discussed at length in case we didn't quite catch it. Humans are storytelling creatures, and even after a complete collapse will still need story, indeed poetry and drama (Shakespeare in this case), and beautiful music—to offer temporary respite. To lift the listener out of their circumstances for a little while—as art has been doing for millennia. The fact that Shakespeare was writing and performing in a time of plague makes St John Mandel's heavy use of Shakespearean material even more poignant. The intertextuality certainly elevates the prose above a predictable dog-eat-dog zero-sum, but is laboured—with the death of the lead actor in a production of King Lear providing the inciting incident, and the fulcrum of the lives touched by it. Arthur Leander, the charismatic and problematic actor has three ex-wives between whom he divides his 'spoils', and experiences his own (emotional) wasteland—even before the world-wide post-pandemic wasteland. St John Mandel focuses on what is essentially a dysfunctional romantic backstory at the expense of much progress in the dramatic present. There are overlong sections which dwell on the 'good old days', and the novel is smothered by its nostalgia for a very Western, middle-class, and (predominantly) white standard of living. The hauntological quality of the post-apocalyptic vistas is very much a love song to late capitalism—shopping malls, burger joints, freeways, and airports, as well as the conveniences of a petite bourgeoise existence. Why else does the author linger so on such scenes which seem to have no real impact on the dramatic present? It seems to reify Jameson's famous dictum: 'It seems to be easier for us today to imagine the thoroughgoing deterioration of the earth and of nature than the breakdown of late capitalism; perhaps that is due to some weakness in our imagination' (1994: xii). There is a longing for normality depicted here that is more than material—St John Mandel seems to hanker for the mainstream consolations of mimetic fiction within a post-apocalyptic piece of speculative fiction. Although the author's moral and aesthetic compass is admirable—and her project here echoes the ethical aesthetics of Fantasy defined as 'Goldendark' (Manwaring, 2018)—the individual components of the novel are unsatisfying, even while the vision is endearing. The apocalypse she imagines is strangely reassuring—there'll be comics! Shakespeare! Classical music! Horticulture! And provincial newspapers!—despite a few inconveniences or obstacles which are no more or less than the daily reality of many living below the poverty line across the world. The conflict is created mainly by a subplot involving a self-styled prophet, a runaway child, and various characters lost in a 'forest somewhere outside (civilisation)'. Like ensorcelled players in a Shakespearean comedy, the members of the Travelling Symphony misplace each other, and just miss each other in the wild—but all's well that ends well. The one section that feels convincingly bleak is the transapocalyptic account of the airport terminal—where survivors arrive and cope with the ensuing stripping away of the old normal. This is impressively imagined and dramatised. It just feels like we have to wade through a lot of middle-of-the-road lifestyle crises to get to it. The fragmented structure may seem ostensibly bold and 'literary', but its dramatic effectiveness is debatable. Nevertheless, there is merit in St John Mandel's dystopian rapprochement—a celebration of simple pleasures, the beauty of the everyday, creative community, and civilised values, which in such times as these is perhaps more important than ever: a vaccine against philistinism and the erosion of decency in popular discourse.

Less optimistic visions of the near-future can be seen in Lauren Beuke's *Zoo City* (2010), set in an alternative version of Johannesburg, South Africa, in which anyone who has committed a crime has an animal familiar magically attached to them thereby creating a mythopoeic apartheid between the 'animalled' and citizens without that atavistic stigma. Although ostensibly similar to Philip Pullman's concept of the daimon in *His Dark Materials* and *The Book of Dust* series—in which every character in his alternate world has an animal companion, one that changes shape throughout infancy, but becomes fixed after adolescence: to be forcibly separated from your daimon is usually fatal, as Lyra's childhood friend, Roger, found. Whileas, in Beuke's novel, the animals are attached in adulthood for transgression against society, not as a totem animal symbolic of their soul, but as a sign of their human carrier's criminal status. And Beuke's 'zoo' is the result of an AIDS-like pandemic, not a sign of 'Dust'—of the innate consciousness of dark matter, the heresy of Pullman's universe.

Pullman's long-awaited follow-up to *His Dark Materials*, *La Belle Sauvage* (2017), is the first in a follow-up trilogy, *The Book of Dust*, and was labelled wittily by Pullman not as a sequel or prequel but an 'equel'. It is entirely set in the alternative world of Lyra Belacqua's Oxford and relates her backstory. Starting in the city, it eventually becomes a chase narrative set on the flooded River Thames—and thus foregrounds the effect of Climate Chaos. La Belle Sauvage (the name of a boat but also possibly referring to at least one of the female characters) has all the thrilling elements that made the first books so great—a vividly imagined storyworld (an alternative England co-existent with this one); convincing, engaging characters (flawed, spirited, with attendant daemons and demons); heightened drama (in the spirit of a Saturday morning serial); a willingness to show pain and darkness (there are shadows here); and a boldness of ambition (once again an engagement with big themes around Religion and Science). Pullman writes for Young Adults with a full-bloodied commitment—a seriousness of endeavour, ideas, and craftsmanship that creates the true 'Crossover' appeal that few Young/New Adult books actually possess, though many claim. While giving younger readers exactly what they would want—young protagonists with agency, a sense of risk and transgression, of empowerment and change, and a gripping read interlaced with the magical and the mundane (food and sleep feature a healthy amount)—he also provides a sense of profundity and subtext, complexity of theme and characterisation, to keep more mature readers on board. Mysterious events happen that are no doubt foreshadowing later elements (perhaps literally) of *The Book of Dust* to be developed. Pullman seems to set things up in a well-paced way in this first instalment—although its sequel, *The Secret Commonwealth* (2019), failed to live up to that promise with more of a conventional spy narrative set aboard a version of the Orient Express. *La Belle Sauvage,* set before the *Northern Lights*, relates Lyra's almost Biblical early years (a special child born upon a flood). It introduces two finely drawn characters, Malcolm and Alice—both are attached to The Trout Inn, a hostelry that actually exists on the edge of Port Meadow, Oxford. But this is the Oxford of Jordan College. Here, Godstow Priory is not a romantic ruin opposite the pub, but inhabited by a gaggle of sympathetically drawn nuns. At this crossroads of the secular and the sacred, dark and light forces gather and worlds are destined to collide. The river (Isis, and downriver, the Thames) dominates the novel—water saturates its pages. The novel's narrative traction is in part created by its inexorable flow, when, after heavy rains, it bursts its banks, and the whole south of England is inundated. Much of the action takes place in the eponymous boat, where Malcolm, Alice, and Lyra attempt to survive a 'rewilded' land, in much the same way as the protagonist of Richard Jefferies' 1885 post-apocalyptic novel, *After London: or Wild England.* Although there are many riches in Jefferies' vivid future primitive scenario, it lacks the narrative thrust of Pullman's flood-world. Pullman's lively prose draws upon the palate of the Gothic and the later Sensationalist fiction of Wilkie Collins et al. As with the previous books the intertextuality (an epigraph from Edmund Spenser's *The Faerie Queene*; a mythopoeic chapter rather like 'The Piper at the Gates of Dawn' from Kenneth Grahame's *The Wind in the Willows*; spy-games straight out of Le Carré) alongside the polished prose bestows upon the novel a classiness that comfortably situates it alongside the works of his fellow Oxford fantastists, Lewis Carroll, C.S. Lewis, and J.R.R. Tolkien, while bringing it 'storming' into the 21st century in a way that foregrounds the Ecological Imaginary.

Another significant work of 21st-century Fantasy which features a flooded environment is Susanna Clarke's *Piranesi* (2020)—a novel dramatically different in tone to both her breakout novel, *Jonathan Strange and Mr Norrell* (2004), and also Pullman's YA fiction. The main setting is a mysterious semi-flooded ruin—the 'House', a vast, architectural world echoing the famous labyrinthine engravings by the titular artist—which seems to be both a psychological space and a pocket universe. The novel is not about that Piranesi (the Italian architect, archaeologist, and artist, 1720–1778), although Giovanni Battista's name is used metonymically in a witty, intertextual way. Clarke's extraordinary Secondary World is a Piranesian prison-world full of vast rooms filled with (almost) innumerable statues, floods, fish, and fowl. Its protagonist, 'Piranesi', meticulously charts its boundless treasures. The only other apparent (living) human is the 'Other', a sartorial psychological manipulator who exploits the friendship and devotion of his unknowing captive. The novel, in its painfully poignant dramatisation of gaslighting, has a more contemporary edge than Clarke's previous works—even before the true nature of the Neoclassical setting is revealed. One could also read an ecological subtext to the depiction of the House's nurturing ecosystem, which Piranesi succeeds in adapting to. The apparently 'man-made' House seems to act as a microcosm for the home ('ekos') of the planet. Yet it also resonates as symbol of the mind—with its various storeys of consciousness, and the way it stores (and misplaces) information. The House seems to homage in particular the 'memory palace' of Simonides. Yet despite Piranesi's scrupulous record-keeping, key things are lost—traumatic memories, identities—things that are telling by their absence, and destabilising of ontologies. In this respect, Clarke's House is a Foucauldian heterotopia. It turns the world (our world) inside out. The usually fey, remote 'otherworld' here is something far more mundane—but this somehow reifies its secular, everyday wonder. As in all good art, Clarke makes us look at the familiar in a new way, and in this regard, it is ultimately life-affirming. Yet it is a bleaker read than her previous two books, and although it may have more of an edge, it is easy to miss the cosier eccentricity and fantastic historicity of them. And Clarke's metanarrative games and use of the haunted house trope (with more of a nod to Emily Dickinson's 'Nature is a haunted house—but Art—is a house that tries to be haunted', than the classic ghost story) would feel familiar to anyone who has read Mark Z. Danielewski's *House of Leaves* (2000), with its Chinese box structure, fake references, and archival material—as well as the disturbing, fathomless, phantom dimensional ghost-house buried within it. It is small wonder that the long-term housebound author has turned an agoraphobic experience into one of such expansive possibility. In this way, *Piranesi* was perhaps the perfect lockdown read—one for all prisoners yearning for emancipation from the walls of reality, and one that has gained an audience beyond Fantasy as a modern-day feminist classic, winning the Women's Prize for Fiction, and loved by a legion of readers (BBC, 2024).

A novel that also has 'escape' of various kinds as a central line of desire is by Pulitzer Prize-winning American author Anthony Doerr. *Cloud Cuckoo Land* (2021) is a storytelling tour-de-force that should appeal to any one in love with fabulation and the fantastic. It is a novel about stories, and a praise-song to libraries, archives, and the endless possibilities of language and the imagination. The multilinear narrative is set over three time periods—Medieval Constantinople, modern-day Idaho, and aboard an intergenerational starship in the not-too-distant future. Each era is focalised through the travails of a small cast of characters: a Christian seamstress and Muslim cattle herder caught on either side of the siege of Constantinople; an ecofanatic malcontent and octogenarian autodidact stuck in a hostage situation in a public library in Lakeport, Idaho; and teenage interstellar native who finds herself aboard a plague-bound starship.

Confinement, restriction, and obligation are explored through these three alternating threads—and linking them all together is an apparently lost Greek epic by Diogenes, about a foolhardy traveller who stumbles into a quest for the titular utopia.

Cloud Cuckoo Lands of different calibres motivate all the characters—from a warm bed and hot meal amid war-torn hardship and displacement; the consolatory refuge of a book; an eco-Valhalla for the hard-line martyr; to an Earth-like planet on the far side of the galaxy. Sehnsucht of varying intensities haunts all of the characters, and its presence rather than its fulfilment seems to be fundamental to the human condition, Doerr seems to suggest. We are motivated by what is absent, what is over the horizon, by the unattainable—so many fool-knights on Grail quests that will never be accomplished. And yet in the striving, in that perpetual, ever-expanding liminal zone between where we are and where we want to be, life is lived. One half riding Baron Munchausen-like to the moon, the other half grounded in the harsh actuality of homegrown terrorism, religious warfare, environmental destruction, the struggles of a single parent raising a child on zero-hour contracts, resource scarcity, and pandemics, Doerr's mercurial novel dances an elegantly fine line between the visionary and the mundane, the didactic and the aesthetic.

Multi-layered, but not dense, its architectural structure is never intrusive: it is cathedral-like in its complexity—that is, it never loses sight of the spire. Everything is perfectly balanced, with this single goal always in mind.

Beyond its technical brilliance, the core of the novel seems to be ecological—in the true sense of the word: it explores and celebrates 'the relation of living organisms to one another and to their physical surroundings' (Oxford Languages). With its long view of the human story, it is a Fantasy novel of the Anthropocene—relating to humans and their impact upon the environment. It hints at the ultimate Cloud Cuckoo Land—a preindustrial, prelapsarian Earth: 'an older and undiluted world, when every barn swallow, every sunset, every storm, pulsed with meaning.' In a way, Doerr's novel attempts to restore that meaning, through an extended act of re-enchantment: by making us appreciate the everyday wonder beneath our feet. It is a novel that lingers in the mind and heart.

A novel set in an ostensibly more conventional Secondary World is *Ombria in Shadow* (2003), although beyond that there is nothing conventional or predictable about Patricia A. McKillip's singular Fantasy. McKillip has created an exquisitely wrought baroque Fantasy in *Ombria in Shadow*, one that builds upon the foundations laid by the Gothic Tradition, as well as that of Medieval Fantasy. A tale of changelings, enchantresses, tragic princes, and dastardly deeds, at the heart of *Ombria in Shadow* is the dark tower of the palace, the centre of all intrigue, tragedy, and magic. There is a tangible sense of place—indeed, the palace is a character in itself, dominating the novel in the way that Domina, the 'Black Pearl' regent dominates the court, city, and country with her Machiavellian machinations. The claustrophobic effect of larger-than-life characters in such a confined space is Gormenghastian. This is compounded by the unusual prose style—McKillip deploys syntactical circumlocutions that curl back on themselves like tangled briars. This could easily be stifling, combining with the Escheresque setting to evoke a density redolent of Umberto Eco's *The Name of the Rose* (1983), or of a Borgesian labyrinth. Yet the effect is more fey and dreamlike, perhaps with a dash of Diana Wynne Jones' *Howl's Moving Castle*. The social hierarchies are certainly codified by the architecture, a pecking order set in stone, but what breathes life into this highly ornate affair (which could easily become a recherché exercise in style) are the great characters—the fallen-from-grace royal consort Lydea, the amoral sorceress Faey and her 'waxling' Mag, the black-hearted regent Domina, and the haunted artist, Ducon. Between them is a tangled web of love, hate, and deception. The court descends into knife-edged chaos when its ruler dies, leaving the only acknowledged heir—a young boy, Kyel—vulnerable and at the mercy of the smothering attentions of Domina. The doubling and mistaken identities that occur throughout are mirrored by the shadow castle that exists, hidden behind walls, secret doors, and beneath the city. This is the demi-monde into which the cast out consort Lydea is plunged—and we follow her down its rabbit hole as she jettisons her courtly attire (or is 'relieved' of it). This foreshadows the stripping away of illusion which must occur for the truth to be revealed, like layers of peeling décor in a mansion of decaying grandeur. McKillip revels in the aesthetic of clutter, squalor, and decrepitude—it is as though the whole city-state is a diseased body which needs purging, a cathartic spring clean to blast away its corruption. However the strength of this novel is less in the plot than in the manner in which it is carried off. There is a febrile energy here, and it is in her depiction of altered states that the author excels. One needs to merely surrender to it and be swept along by its fever dream.

Of a similar ilk, in terms of the prose, and the setting, is Jeannette Ng's *Under the Pendulum Sun* (2017). There is much to commend in Ng's 'novel of the Fae' about troubled sibling missionaries, Catherine and Laon Helstone, and their strange adventures in Elphane/Arcadia. Ng manages to evoke through her finely wrought prose the claustrophobic atmosphere of a dense Victorian novel of morality and misadventure; and also the alien, quixotic climate of Faerie. By making her brother and sister protagonists missionaries seeking out to bring the Word of the Lord to the benighted 'souls' of this recently revealed Otherworld the novel both aligns with and subverts the colonial project—for the 'too close' missionaries are far from without sin, and their mission is futile at worst, at best a metaphysical challenge (do the Fae have souls? what is their place in God's creation?). Access to the 'inner lands' for further proselytising is the main plot McGuffin, but the chief line of desire revolves around Catherine's unhealthy obsession with her brother—who is a Branwell/Heathcliff/Rochester type. Dark, moody, and (to some) irresistible. This is not surprising as Ng clearly riffs upon the Brontë family dynamic and legendarium (which the famous siblings of Haworth created in their younger years). Here, this juvenilia is given the full-bloodied treatment, as Ng feeds it into the mulch of her world-building. The mise-en-scène of each chapter is vividly imagined, but often this seems to be at the expense of narrative traction. Sometimes it is hard to know exactly what is happening—many of the scenes have the feverish intensity and illogic of a dream. And although the minutiae of Elphane, in particular life in Gethsemane, the Pale Queen's castle, is exquisitely imagined, the broader brushstrokes of this Secondary World are less convincing: the pendulum sun of the title, the fish of the moon swimming in the sky, and sea whales (which seem to be made of rickety whicker, yet contain a microcosmic ocean). This no doubt is intended to deliberately subvert the verisimilitude and make the otherworldly realm lack naturalism—and such bold imagery may be original and memorable, it threatens to make the whole edifice a leaky vessel. Yet unlike a lot of (modern) fantasy, Ng's prose aspires to a slightly elevated register, which successfully evokes the music of strangeness ('a catch of the breath', as Susan Cooper describes it). Ng's depiction of Faerie is one of the best seen in contemporary fantasy. She lards each chapter with an epigraph, pastiches written wittily in the style of bombastic Victoriana, or stuffy exegeses. These often evoke the texture of an A.S. Byatt novel (notably her 1990 Booker Prize-winning *Possession*) but are convincingly done. Ng's academic background and interests (MA in Medieval and Renaissance Studies/Medieval and Missionary Theology) clearly inform these, but they can risk becoming rather laborious after a while (one can always choose whether to read them or not). Perhaps too much salt, and not enough meat for my taste.

A Fantasy novel that interrogates the fault-lines of colonialism more explicitly is *Babel, or The Necessity of Violence* (2022). *Babel* is a dark academia novel by R.F. Kuang, author of *The Poppy War* trilogy, and a dazzling academic herself: a Marshall Scholar and Chinese-English translator, with an MPhil from Cambridge and an MSc from Oxford, and later engaged in doctoral study at Yale. This (possibly) standalone novel concerns a talented young Cantonese boy who adopts the name Robin Swift. He is plucked from the deathbed of his mother amongst the slums of his homeland by a strict Oxford professor, who cultivates his ward's gift with languages to the point where Swift is able to become a student of the prestigious Royal Institute of Translation—the tallest building in the heart of an alternative Oxford. Here, the power of translation is used to activate silver bars which generate the qualities evoked by the 'matching pairs' of the translated languages: energy, wealth, control. These drive the dominance of the British Empire, growing fat and powerful via the exploitation of the resources, slave labour, and talent of their 'subjects' from the colonies and from other countries.

At first Swift is enthralled by the City of Dreaming Spires, but soon learns the harsh realities of being non-white and non-Western, however talented. Fortunately he finds kindred spirits in a small group of friends who quickly bond, united by their outsider status: Black; Asian; Chinese; female; Muslim—all reasons for their fellow students, townsfolk, and the institution itself to ostracise them.

Yet when Swift stumbles upon a burglary of silver bars and instinctively helps someone who appears to be his twin, he becomes embroiled in a dangerous demi-monde, where a secret society fights to overthrow the rapacious Empire. Kuang has clearly drawn upon her experience of being a student at Oxford, even though she is at pains in her foreword to emphasise it is only 'a work of fiction'.

Nevertheless, much of the novel's impact comes from not only its vivid sense of place (in a slightly tilted Oxford reminiscent of Pullman's Jordan College) but an excoriating deconstruction of the glittering shibboleths of her alma mater—its elitist snobbery, territoriality, and exploitative neoliberal model, which hoovers up and capitalises upon talent.

Most of all, and perhaps more mercilessly and insightfully, it exposes the Colonialist and Racist foundations of the university, and of the British Empire, for which it serves as a metonym. In a paratextual way, the footnotes provide a counter-narrative to this hegemony. Many are factual, perhaps belying Kuang's scholarly habits, but some are counterfactual, in the spirit of Clarke's *Jonathan Strange and Mr Norrell* (2004).

A Fantasy novel about language is going to be riddled with intertextuality, and that is the case here. It is not only a love letter to language, to libraries, and to global literature in particular, but also a cri-de-coeur against its weaponisation and commercialisation. Polyglottal fluency both empowers and enslaves the protagonists: they become 'assets' of the Empire, but ones with agency who choose to rebel.

The friendship between the four of them, the nuanced and fluid dynamics, is brilliantly dramatised—their fellowship provides the cornerstones of this towering novel, whose vaulting ambition is veritably Promethean. It seems to want to tear down the whole corrupt edifice of Western 'civilisation', and evoke revolt—a poignant coda of Percy Shelley's 'The Masque of Anarchy' perhaps (Shelley famously being expelled from Oxford for his paper on atheism).

As a late member of the loose circle of writers known as the Oxford Fantasists (whose members include Lewis Carroll, C.S. Lewis, J.R.R. Tolkien, Susan Cooper, and Philip Pullman), Kuang seems determined to turn the burning glass of the imagination on the city itself. Here, the Fantasy feels closer to home and nearer to the knuckle, exposing painful home truths.

Yet, the novel never becomes a mere diatribe or screed, thanks to Kuang's command of the narrative. The chapters are as solidly constructed as chambers in the titular tower—with everything leading to the spectacular conclusion, with only the odd indulgent digression for scones: a self-confessed weakness of the novelist's, and a couple of side-chapters, fleshing out two of the supporting characters. A magical education narrative could easily become Potterish in lesser hands, but the bite of realism and constant reminders of the fault-lines of society help to avoid any cosy supernaturalism. Kuang goes for the jugular. This is a finely controlled, but fierce, roar of a novel, an Oxonian *Fight Club,* inflected by the emotive debates that have broken upon the walls of academe recently around colonialism, the legacy of slavery, equality, diversity, and representation.

Babel (2022) could be said to explore the Ecological Imaginary through its charting of the economic forces which drive global trade, and how that impacts upon lives far and wide in a profound, and often decimating way. Kuang seems to be saying there is no separation, and there is no escaping the yoke of complicity and culpability.

The Dark Fantastic

No discussion of 21st-century Fantasy can ignore the impact and importance of the greater diversification of writers, and better representation of characters from the BIPOC and LGBTQI+ communities. In Fantastika, authors write not only across gender, ethnicity, and class, but across worlds. Even if the invention of other orders of being in a Secondary World seems liberating, it still comes with responsibility. The Fantasy author is not let off the hook, but must approach their 'sub-creation' with skill, compassion, and mindfulness. Representation must avoid stereotyping, tokenism, and essentialism, for such failures of imagination and craft affect the non-diegetic discourse of the novel, and how it is received by reviewers and readers.

Largely thanks to the Black Lives Matter movement (founded in 2013 by Alicia Garza, Patrisse Cullors, and Opal Tometi), publishers realised they had a crisis of equality and representation that needed to be addressed, and began to diversify their lists, commission work from under-represented and marginalised communities, and actively promote writers of colour. Science Fiction had made a head start with the work of Samuel R Delaney, Nalo Hopkinson, and Octavia Butler. As Nalo Hopkinson said in her keynote speech at 'Once and Future Worlds' in Glasgow, 2022: 'If you don't fit the market, make the market expand to fit you.' Yet in Fantasy, BIPOC writers were, at least until the early 21st century, poorly represented with very few exceptions. That is not to say Fantasy was not read and enjoyed by members of the BIPOC communities around the world, but as Marlon James pointed out, it was harder to relate to stories where you did not see people like yourself represented. Instead, James initially found a deeper connection through superhero comics, especially the outcasts of *The Uncanny X-Men*, partly because 'comic books were easier to get hold of than books' in his childhood home of Jamaica (*Guardian*, 2019). He wrote and drew his own version of a group of super-powered misfits who come together to fight crime, before attempting his first novel: a fantasy novel. Yet as a mature, prize-winning writer of Booker Prize-winning 'literary' fiction, James felt he did not want to be compelled to write only upon certain subjects, or in certain genres: 'As people of colour in the diaspora, we're particularly interested in stories that go beyond slavery. I'm tired of that being seen as the furthest in the past we can go or that swords and sorcery aren't available to us' (*Guardian*, 2019).

A step in the right direction in the wider umbrella of Fantastika was the formulation of 'Afrofuturism' in the mid-90s. Coined by Mark Dery in 1994, *Black to the Future*, although with roots going back to the psychedelic-tinged jazz of American artiste Sun-Ra and Marvel's *Black Panther*, Afrofuturism is a fluid, and sometimes contested, term—Nnedi Okorafor prefers 'Africanfuturism' (Ogbunu, 2023)—but has been defined by Ingrid LaFleur as: '. . . a way of imagining possible futures through a black cultural lens. I see Afrofuturism as a way to encourage experimentation, reimagine identities and activate liberation' (TEDx talk). Ytasha Womack explored this more deeply in her book: 'Afrofuturism is often an amalgamation of narratives, but at the core, it values the power of creativity and imagination to reinvigorate culture and transcend social limitations' (2013: 24). Womack acknowledges that '[w]ith the diversity of the nation and world increasingly standing in stark contrast to the diversity in futuristic works, it's no surprise that Afrofuturism emerged' (2013: 6). Honouring the influences of the subgenre, Womack sees the larger imaginary space it affords: 'the creativity born from rooting black culture in sci-fi and fantasy is an exciting evolution' (2013: 10). She claims that 'Afrofuturism unchains the mind' (2013: 15) and sums it up as "The predominance of soul culture gone cyberpop' (2013: 2). Critically, she observes the empowering, self-defining quality of Afrofuturism, 'Afrofuturists write their own stories' (2013: 16). She goes onto observe the radical, activistic quality to Afrofuturism: 'being imaginative and creative, and even projecting black culture into the future, was part of a lineage of resistance to daunting power structure' (2013: 19). Womack draws parallels between the slave trade and alien abductions, arguing that it is not surprising writers of colour can easily imagine characters transported to other realities (being themselves the legacy of colonial extraction and exploitation), or are keen to imagine counterfactual narratives such as Colson Whitehead's Pulitzer Prize-winning *The Underground Railroad* (2016), reclaiming the past in what has been referred to as 'Steamfunk', that is black steampunk. For a sense of the breadth of work being generated under this loose banner, see the 2022 exhibition catalogue, 'In the Black Fantastic', based upon the exhibition at the Hayward Gallery, London, curated by Ekow Eshun (2022). And partly as a result of this wonderfully rich new wave of Fantastika, the bookshelves and screens have been enriched by titles such as N.K. Jemisin's *Broken Earth* trilogy, *The Fifth Season* (2015), *The Obelisk Gate* (2016), and *The Stone Sky* (2017); Nnedi Okorafor's *Lagoon* (2014) and *Who Fears Death* (2010), *Black Wolf, Red Leopard* (2019) and *Moon Witch, Spider King* (2022), *Black Sun* (2020); Tade Thompson's 'Wormwood' trilogy, *Rosewater* (2016), *The Rosewater Insurrection* (2019), and *The Rosewater Redemption* (2019); Fonda Lee's *Jade City* (2017), *Jade War* (2019), and *Jade Legacy* (2021); Tasha Suri's *The Jasmine Throne* (2021), and S.A. Chakraborty's Daevabad Trilogy, *The City of Brass* (2017), *The Kingdom of Copper* (2019), and *The Empire of Gold* (2021); Premee Mohamed's *'Beneath the Rising'* trilogy, *Beneath the Rising* (2020), *The Broken Darkness* (2021), *and The Void Ascendant* (2022); Sofia Samatar's *A Stranger in Olondria* (2014); Yoon Ha Lee's *Ninefox Gambit* (2016); Victor Lavalle's *The Ballad of Black Tom* (2016) and *The Changeling* (2018); Kacen Callender's *Queen of the Conquered* (2020); R.F. Kuang's *The Poppy War* (2018); Nalo Hopkinson's *The Salt Roads* (2004); Saladin Ahmed's *Throne of the Crescent Moon* (2013); *Ken Liu's The Grace of Kings* (2015); Silvia Moreno-Garcia's *Mexican Gothic* (2020); Chloe Gong's *These Violent Delights* and *These Violent Ends* (2020); Shelley Parker-Chan's duology, *She Who Became the Sun* (2021) and *He Who Drowned the World* (2023); Zen Cho's *Sorcerer to the Crown* (2016*)* and *The True Queen* (2019); Aliette de Bodard's *The House of Shattered Wings* (2015); Ta-Nehisi Coates' *The Water Dancer* (2020); Andrea Hairston's *Redwood and Wildfire* (2011); G. Willow Wilson's *Alif the Unseen* (2012); and P. Djeli Clark's *A Master of Djinn* (2021) to name a few.

This strand of Fantasy—most welcome and much-needed—has been discussed in depth by Ebony Elizabeth Thomas (2019), Attebery (2022), Sangster (2023), and others, and is beyond the scope of this current book to discuss, as the focus is primarily on the Ecological Imaginary. Even if such books do not engage directly with this—with notable exceptions duly acknowledged—*all* are inflected by Bould's notion of the 'Anthropocene Unconscious' (2021). Bould boldly posits that *all* texts created now (in the 21st century) belie trace elements of the time they are written in: the Anthropocene. They all (will) create a telling layer in the stratigraphic record, a dark stain that speaks of petroculture's endgame: 'The art and literature of our time is pregnant with catastrophe, with weather and water, wildness and weirdness' (2021: 3) and 'the Anthropocene *is* the unconscious of 'the art and literature of our time' (2021: 15). With a pancultural largesse he sees any kind of text, however 'low brow' as fair game—a stance which he demonstrates by analysing the 'Sharknado' film series, and *The Fast and the Furious* franchise, alongside the likes of Paul Auster, Richard Powers, Thomas Hardy, Amitav Ghosh, and Arundhati Roy. Bould argues for entanglement, for complexity, for a 360-degree inclusiveness and criticality, for cutting the academic jabber-jabber, for going for the jugular:

> But down here in the Fast-and-Furiocene, our choice is clear.
> Ecosocialism or barbarism.
> Ride or die.'
>
> (2021: 142)

His prose and his thesis is provocative, but suggests the urgency and importance of what we face, blasting away the complacency and quietism of *any* cultural artefact that suggests it should be 'business as usual'.

* * *

No doubt, the exponential growth in the development of and use of Generative AI will have an impact on 21st-century Fantasy (and more widely and devastatingly on publishing in general). The ethical ramifications have been much discussed in Posthuman Studies, but the environmental factors are rarely considered (e.g. the huge use of precious water resources to maintain server farms).

How the rise of the Far Right will impact not just Fantasy, but society in general, also waits to be seen.

6 Mapping the Fantastic

The Ethics of Placiality, Architectonics, and Speculative Cartography

Unfolding the Map

Does every Fantasy begin with a map, or at least with the desire to 'map', to explore, to know, to record, and to share? Tolkien would perhaps concur, saying in a BBC interview: 'If you're going to have a complicated story you must work to a map, otherwise you'll never make a map of it afterwards' (1964). It is tempting to suggest a basic demarcation of Fantasy authors into those who include maps in their work, and those who do not: mappers; and non-mappers (although sometimes maps are added to a Fantasy text posthumously, or embellished through various iterations). The latter camp would perhaps find affinity with the following challenge to this cartographic tendency: do maps foreclose on Fantasy, colonising our imagination before we've even started reading? Despite Pratchett's wry note in lieu of a formulaic Fantasy map: 'You can't map a sense of humour,' (1983) from the first in his Discworld novels, a series that became obsessively mapped—many readers and increasingly fans or 'prosumers', to use Toffler's term (1980) of Fantasy, would no doubt disagree with the notion that a map is counter-productive to the very endeavour of Fantasy, seeing the map as an *intrinsic* part of the enjoyment and engagement with the storyworld, as Lewis-Jones has pointed out in his broad survey of 'Imaginary Lands': 'As Tolkien realised, the literary map serves both to guide the author and to make the reader feel a part of the expedition' (2018: 68). And, regardless of one's position on the inclusion or exclusion of the Fantasy map within the paratext of a narrative, is not any attempt to 'map' this terrain theoretically just another kind of mapping—one that some writers of Fantasy either dislike or actively resist? These are some of the questions I shall explore here.

* * *

Every setting in a story suggests its own vectors of ethicality—a castle with its throne-room and dungeons; a city with its palaces and slums; a floating utopian island (and all *that* excludes); a pirate cutter; a caravanserai; or a fairy realm. World-building always implies various inter-related ethical

DOI: 10.4324/9781003463665-7

choices—economic systems, magical systems, and ecosystems—like the lines of directionality that once used to criss-cross old charts. The charting of such a world involves numerous acts of selection, omission, and hierarchy, as Reif Larsen suggests: 'As writers we can never include it all, but like magicians, we can erect enough scenery so as to give the illusion that we are including it all, and our readers will fill in the rest' (cited in Lewis-Jones, 2018: 169). These intentional negative spaces are seductive elements within any narrative. Peter Turchi, in his exploration of the writer as cartographer posits: 'omissions, intended or unintended, provoke the imagination' (2004: 47). Thomas Wolfe, in a letter to F. Scott Fitzgerald, summarised this pithily: 'A great writer is not only a leaver-outer but also a putter-inner' (cited by Turchi, 2004: 41). Seeing such decisions as intrinsic to the creative process, and the development of art, Turchi draws an extended comparison between cartography and the act of composition: 'A story or novel is a kind of map, because, like a map, it is not a world, but it evokes one (or at least one for each reader)' (2004: 166).

What is central on the map, and what is marginal? Every capitol implies the provincial, every crossroads of civilisation the wilder edges where dwell dragons. Geology, geography, and the geopolitical co-exist cartographically within the same lines of longitude and latitude, especially in the metaphoric realms of Fantasy, where difference and division are often personified in person and place.

In this chapter I explore the ethics of architecture and 'placiality' (Casey, 1988) as a way of acknowledging that the depiction of 'nature' and 'landscape', or any environment, is never neutral. Within the construction of Fantasy what are the intrinsic ethical fields implied by the 'power of place', as geographer Tim Cresswell calls it (2014: 6)? The historian Simon Schama said of 'landscape', the creation and perception of which is a form of mapping: 'Before it can ever be a repose for the senses, landscape is a work of the mind. Its scenery is built up as much from strata of memory as from layers of rock' (1995). These strata are sometimes hard to perceive, especially as the observer is implicated in what is beheld—a 'landscape', or the representation of a landscape in cartographic, artistic, or literary form. How to perceive clearly not only what is 'there', but what is telling by its absence is the task of any critic, who are involved in their own, very specific, kind of mapping: their projections just as speculative as any Mappa Mundi, as distortional as Mercator's or Peter's.

And yet, 'map' we must. Nature, it is said, abhors a vacuum; the attentive reader leaps upon it; and the scholar and critic savours it.

Fundamental axes of such considerations—the eastings and northings of our purview here—would be perhaps 'place' and its valuation. How significant is a place to the narrative, and what are the implications of that significance? Although all such places within a Fantasy narrative landscape could suggest by their importance, or lack of, valuation of some kind, it is clear that certain landmarks or locales dominate, and these inevitably draw our gaze, our imagination and interest, and act as focal points for ethical consideration—and thus will focalise our discussion here. Nevertheless, it is acknowledged that for every Gormenghast, Mount Doom, Hogwarts, Ankh-Morpork, or Lud-in-the-Mist, there are places that are unrecorded or unexplored: the 'empty' places on the map, or off the edges. And inevitably, in a mapping of any territory, selection and omission will occur. It is unworkable, within the parameters here, to cite every example. This brief survey will undoubtedly be subjectivised by the inherent interests and unconscious bias of its surveyor, but will hopefully give enough of an indicator of each area to prompt discussion.

Walter Benjamin suggested that [w]ork on good prose has three steps: a musical stage when it is composed, an architectonic one when it is built, and a textile one when it is woven' (1928). In the architectonic, I consider both the structural composition of a novel and the structures contained within it.

The cartographic aspect has already been mentioned, and underpins the whole enquiry, and the 'textile' has been discussed throughout the book in the analyses of the creative process of Fantasy texts.

I will begin with a survey of the critical terrain, before zooming in on key vectors: placiality, planetarity, and the Dark Fantastic. Then I will consider more radical approaches: autocartography and non-mapping, post-colonial counter-mapping, and the regendering of space. In the penultimate section I will consider the argument 'against' world-building, before summarising the ground covered and identifying critical points.

The Critical Terrain

Critical theories are frames for viewing a certain territory—a particular discipline, author, oeuvre, or genre (to list a few possible areas of focus)—although sometimes it seems they risk becoming the territory itself, obscuring or even replacing that which is discussed, as in the 'Grand Map' mooted in Lewis Carroll's late novel, *Sylvia and Bruno Concluded*:

> "And then came the grandest idea of all! We actually made a map of the country, on the scale of a mile to the mile!"
>
> "Have you used it much?" I enquired.
>
> "It has never been spread out, yet," said Mein Herr. "The farmers objected: they said it would cover the whole country, and shut out the sunlight! So we now use the country itself, as its own map, and I assure you it does nearly as well."
>
> (Carroll, 1894)

Mein Herr's idea, the epitome of Carroll's queasy logical inversions, but smacking of the Late Victorian obsession for omniscience—with its exhaustive cataloguing and categorising and desire for pink-mapped Imperial dominance—was critiqued by Borges in his darker, post-war variant, 'On Exactitude [or Rigour] in Science' (1946): an Ozymandian vision given a layer of faux-historicity by masquerading as a quote from a 17th-century source:

> In that Empire, the Art of Cartography attained such Perfection that the map of a single Province occupied the entirety of a City, and the map of the Empire, the entirety of a Province. In time, those Unconscionable Maps no longer satisfied, and the Cartographers Guilds struck a Map of the Empire whose size was that of the Empire, and which coincided point for point with it. The following Generations, who were not so fond of the Study of Cartography as their Forebears had been, saw that that vast map was Useless, and not without some Pitilessness was it, that they delivered it up to the Inclemencies of Sun and Winters. In the Deserts of the West, still today, there are Tattered Ruins of that Map, inhabited by Animals and Beggars; in all the Land there is no other Relic of the Disciplines of Geography.
>
> —Suarez Miranda, Viajes de varones prudentes, Libro IV, Cap. XLV, Lerida, 1658

With the advent of civilian-access satellite mapping technology, such as Google Earth, the Grand Map would appear to be a tangible reality (Steinberg, 2011), or at least one at our fingertips, and yet its inexhaustibility is illusory. A map that was truly 'complete' would quickly become redundant, as Turchi pointed out: 'Whatever a map's attitude towards blanks within its borders, virtually everything is left off a map—and must be for a map to be useful' (2004: 40). An observation echoed by Denis Wood: 'No map can show everything. Could it, it would no more than reproduce the world, which, without the map, we already have. It is only its selection from the world's overwhelming richness that justifies the map' (cited by Turchi, 2004: 40). A map is never more than a highly selective representation, the creation and design of which is highly predicated upon the ethos of its cartographer (or commissioner). And yet it is all too common to mistake the map for the territory, as Neil Gaiman wittily points out, via the notebook of Mr Ibis, in *American Gods*:

> One describes a tale best by telling the tale. You see? The way one describes a story, to oneself or to the world, is by telling the story. It is a balancing act and it is a dream. The more accurate the map, the more

> it resembles the territory. The most accurate map possible would be the territory, and thus would be perfectly accurate and perfectly useless.
>
> The tale is the map which is the territory.
>
> You must remember this.
>
> (Gaiman, [2001] 2004: 589)

Despite Mr Ibis' reminder that the *tale* is the map itself, defining the territory *through the telling*, it is all too easy for the actual source texts—the humble Fantasy novel—to be subsumed by the seemingly endless exegeses that comment, critique, and frame them: a whole academic sub-industry in itself (which this current work is a part of). One could create a meta-map of Fantasy, based upon the cultural theory and critical discourse that has accreted over the 'original' layer of the respective ur-text (often a palimpsest in itself of earlier textual influences), like an ancient civilisation obscured by later ones. Visualise a Rorschach inkblot of continents. First, note the structuralist village of Propp. The comparative crossroads of Campbell. Beyond, behold the Kingdoms of Attebery, Mendlesohn, and Clute-Grant; the twin-cities of Jackson and Hume; the ethical frames of Altes, Phelan, Booth, and Shen; the citadel of Todorov; the Forest of LeFebvre; the Nomad Space of Deleuze and Guattari; the Free Republic of Walter Benjamin; the Pirate Utopia of Hakim Bey; the Oriental Empire of Said; the Global South of Spivak; and the Dark Fantastic continent of Thomas (or rather the mantle, for the Dark Fantastic is a concept that touches every corner of Fantasy's imaginarium), to identify a few features. Any such map has finite parameters, telling absences, and zones of terra incognita—yet around the edges of our map, we may find traces of other transdisciplinary 'charting' (Cresswell, 2014; Buck-Morss, 1991). Hidden within the compass rose, we may discover a Guild of Cartographers featuring Turchi, Lewis-Jones, and other cartophile scholars. At an urban, granular level, we may spot the free radicals of psychogeographers (Debord, 1954; Sinclair, 1997, 2003; Ackroyd, 2000; Coverley, 2018; Solnit, 2001; Elkin, 2016), mythogeographers (Smith, 2010), and their ilk, obsessively charting the hyper-local. If we look even closer, we may discern the fine, mycelial networks of mycologists (Rayner, 2010; Sheldrake, 2020), creating a corpus callosum across the hemispheres of the world (2022). At a macro level, perhaps depicted as a blustery god of weather, Lovelock offers to frame the world as a self-regulating biosphere: Gaia (first proposed by Margulis and Lovelock in 1973). The map is, by definition, never finished, never complete. Waves of scholars continue to add to it, identifying gaps, or insisting upon corrections. In increasingly small handwriting they add addendum to the existing text. Sometimes this language itself is deconstructed, as in the ecolinguistic analysis expounded by Stibbe (2021). Occasionally, Fantasy authors themselves help to remap the territory—deconstructing the implicit and explicit colonialism, gendering, and other biases of their antecedents. This is being pioneered by a panoply of brilliant writers of colour—Nalo Hopkinson, N.K. Jemisin, Marlon James, Nnedi Okorafor, Jeanette Ng, S.A. Chakraborty, Fonda Lee, to name only a few. Critics, too, are helping to remap Fantasy, for example, Mark Dery's 'Afrofuturism' (1993); Mendlesohn's useful 'non-taxonomy' (2008); Palmer-Patel, with her reconsideration of the Campbellian 'Hero's Journey' in American fantasy (2020); Dimitra Fimi, with her much-needed study on Tolkien, race, and cultural history (2009); Guanio-Uluru's analysis of ethics and form in Tolkien, Meyer, and Rowling (2015), Casey's 'placiality' (1998), and Ekman's survey of the Fantasy map (2013). All such 'mapping' has its use, and will be drawn upon accordingly, but never at the expense of the primary sources themselves, to which we shall accordingly turn. But, first, I would like to consider some key vectors in what could be called, at a macro and micro level, 'ethical mapping', and 'conscionable setting'.

Key Vectors: Placiality, Planetarity, and the Dark Fantastic

Firstly, a key term in such an endeavour is Edward Casey's notion of 'placiality' (1998). David Kolb sums up Casey's concept:

> Casey's overall idea is that place is not simple location in uniform geometrical space, but concrete bodily locatedness. Place is qualitative, relational and textured, enclosing and defining. The opening of place happens in the intertwining of nature and the lived body.
>
> (Kolb, 2022)

Casey emphasises the fundamental importance of place, declaring: 'we are implaced beings to begin with, that place is an a priori of our existence on earth' (1988: x). Resisting the colonising of 'place' by 'space', which Casey memorably describes as 'a cosmic and extracosmic Moloch that consumes every corpuscle of place to be found within its greedy reach' (1988). This absorption into the wider term has resulted in 'place' being relegated to mere 'location'. Casey takes a phenomenological approach in his reclamation of the term. Stephen Hardy, in his analysis of Casey's concept (2000), carefully parses the concepts of 'space' and 'place'—acknowledging that, although the latter could be considered a subcategory of the former, our relationship with it is qualitatively different. As Casey himself puts it: 'Place is a part of space which a body takes up' (1998: 150). Following in the footsteps of Husserl, Heidegger, Merleau-Ponty, and Deleuze, Casey sees place as something that is experienced in a proprioceptive, multi-layered way: what could be called the 'affect'. This suggests a place in Fantasy is never merely a 'location', but the convergence of several discourses:

history, memory, folklore, class, gender, ecology, the more-than-human, and so forth. The concept of placiality heightens our sensitivity to the use of setting within a Fantasy narrative. A setting is never a mere backdrop, one to only lightly sketch in if at all. It is a critical part of the story, and key in establishing tone and theme. Indeed, for many readers it is setting that draws them to a particular kind of book: be it a sylvan setting (e.g. Holdstock's *Mythago Wood* cycle), islands (Le Guin's *Earthsea Quartet*), tall ships and high seas (Barker's *Tide Child* trilogy), an otherworldly city (Chakraborty's *Daevabad*), fairy court (Sarah J. Maas' *A Court of Thorns and Roses* series) or somewhere closer to home (Garner's Alderley Edge sequence). For many, Fantasy is a 'place' they escape to, within which are contained myriad 'places'—some of which appeal, some repulse—locales and locations through which vicariously we explore. Some of these get reified through fan-fic, art, computer games, models, theme parks, and tours of TV and film locations. And thus, placiality is a key concept in the re-imagining of setting in Fantasy.

Another key vector in the creation of an ethical setting is a deep awareness of the bigger picture, and avoidance of the tendency to marginalise or exoticise, to skew things towards the Global North, or the Eurocentric, or medievalism—any hegemonic 'default'. Spivak offers a useful concept to epitomise this more outward-looking consciousness: 'Planetarity':

> The planet is in the species of alterity, belonging to another system; and yet we inhabit it, on loan. It is not really amenable to a neat contrast with the globe. I cannot say 'the planet, on the other hand.' When I invoke the planet, I think of the effort required to figure the (im)possibility of this underived intuition . . . . Planet-thought opens up to embrace an inexhaustible taxonomy of such names, including but not identical with the whole range of human universals . . . . If we imagine ourselves as planetary creatures rather than global entities, alterity remains underived from us; it is not our dialectical negation, it contains us as much as it flings us away.
>
> (2017)

Spivak is at pains to disassociate this from globalisation, which she critiques as 'the imposition of the same system of exchange everywhere'. In direct resistance to this default capitalist discourse, she sums up Planetarity as 'socialism at its best': it is communality on a planet-wide scale. Yet, beyond the Marxism, such a concept offers a sympathetic frame for alterity: for a world where we can hear the 'subaltern' speak. This is directly applicable to the creation of Fantasy worlds. In the laziest of world-building, it is easy to relegate wildlings and 'savage' tribes to the marginal Badlands—uncivilised and, critically, unproductive—a useful myth for conquering nations perpetuated since Caesar's *Conquest of Gaul* (1982).

To counter such myths, it is also useful to consider indigenous perspectives, as for example shared and extrapolated upon by the scientist Robin Wall Kimmerer, who shows how the stories and plant wisdom of First Nations people can offer guidance for living in harmony with the biosphere at a micro and macro level (2020 [2013]). Deep ecologist Joanna Macy describes the benefits of this approach: 'Action on behalf of life transforms. Because the relationship between self and the world is reciprocal, it is not a question of first getting enlightened or saved and then acting. As we work to heal the earth, the earth heals us' (cited by Wall Kimmerer, 2013: 340). Animist societies are a common trope in Fantasy, and one could argue that—traditionally—its literature has been predicated upon a supernatural paradigm, one of consequence and exchange between 'worlds', in the way that its younger sibling, SF, is contingent upon a scientific paradigm. Of course, there are many exceptions to this, especially in hybrid forms such as 'Science Fantasy'. But the animist trope is pervasive—and would now have to withstand the criticism of cultural appropriation. However, a fictional tribe based upon a specific culture is arguably better than one that is a generic 'native' tribe. Exactly how it is depicted, the level of research and sensitivity in its rendering, can transform such a questionable cliché into something redeemable. Intentionality and sophistication make all the difference. And yet some have argued that only writers from that particular culture have the right to tell their story (Keene, 2015); others have resisted it but often with diametrically opposed agendas (Shriver, 2016; Evaristo, 2019); others have taken a more nuanced approach, such as Kit de Waal:

> As writers we have to be the other—without it we would have no literature, no great stories, no murder mysteries, no great romances, no historical novels, no science fiction, no fantasy—but when we become the other we need always to act with respect and recognise the value of what we discover, show by our attitudes and our acknowledgements that we aren't just appropriating but are seeking to understand.
>
> (2018)

In Fantasy this is complicated by the fact that direct representation is relatively rare, and amalgamations and derivations are more likely. Nevertheless, the emergence of indigenous voices in Fantasy, such as Rebecca Roanhorse, is a welcome corrective to the problematic 'ventriloquising' of predominantly white authors. As Nnedi Okorafor is keen to point out: 'Of course, anybody can write about anything. Nobody wants to shut anything down. But people [of a given community] should tell their stories first. Because there are things that cannot be captured by outsiders. Some stories are just meant to be told by certain people' (2016). This echoes the Ghanaian proverb: 'Until the lion learns to write, the hunter will always tell the story of the hunt.' A spectrum of perspectives is always better than a monolithic monologue, but it is vital that such voices are no longer silenced.

Thomas's conceptualisation of the 'Dark Fantastic' (2019) argues for a conscious recalibration of the whole Fantasy landscape. It was driven by a desire to write stories that acknowledged cultures beyond the Eurocentric: 'As a young writer, I audaciously dreamed about someday creating and publishing my own fantasy stories that weren't based primarily on European folklore or myth' (2019: 16). Drawing upon Cohen's 'Monster Theory' (1996), Thomas advocates an unsilencing of what she calls 'Dark Others': 'Critical race counterstorytelling provides both translation and amplification for these subsumed narratives' ('Monster Theory', 1996: 23). Thomas wishes to bring the 'Dark Other' into the centre of the Fantastic narrative, not exorcise it, but by transforming the eternally marginalised 'Dark Other' into the 'Dark One' ('the subject, the focalizer, and the narrator' 'Monster Theory', 1996: 25). To jettison it is to render the Fantastic impotent and anaemic. She emphasises, 'The Dark Other is the engine that drives the fantastic' ('Monster Theory', 1996: 25). She believes that 'there can be no fantastic without the Dark Other. The Dark Other is the counterbalance, the counterweight that makes the entire enterprise of the fantastic work. Even when the Dark Other is not present, her present-absence nonetheless haunts the story' ('Monster Theory', 1996: 28). But this is not achieved without effort and skill, and, Thomas argues, a whole new way of thinking about Fantasy: 'Liberating the fantastic from its fear and loathing of darkness and Dark Others not only requires new narratives for the sake of endarkened readers. It requires emancipating imagination itself' ('Monster Theory', 1996: 29). Ultimately, Thomas sees this as an essential project not just for the genre—it has wider implications: 'Working toward a fantastic that is restorative, transformative, and emancipatory has the power to remake the world' ('Monster Theory', 1996: 13).

Such bold conceptualisations can sometimes be world-changing, and the final two ideas I wish to consider here have done just that—they are truly paradigm-shifting, even if they remain contentious. The first is James Lovelock's contested 'Gaia' hypothesis, first mooted in a paper co-authored with American microbiologist Lynn Margulis in 1973. It postulates that 'all living things have a regulatory effect on the Earth's environment that promotes life overall; the Earth is homeostatic in support of life-sustaining conditions' (2022). When it was published more widely (1995, 2000, 2009), it caught the public imagination at a time when there was growing concern about the environment due to the increasingly discernible effects of Climate Change, the transnational climate agreements of Rio 1992 onwards, and the reports of the IPCC. Drawing upon Greek mythology to provide a personification of his concept, it is not surprising that such a ground-breaking idea can feed back into the worlds of Fantasy, which often draws upon not only mythology but also anthropological (Le Guin), economic (Pratchett), and scientific (Pullman) theories. A world that is conscious, self-regulating, that has agency—is in fact, a 'character' in its own right—is one that contemporary Fantasy authors have embraced for example, Holdstock in his *Mythago Wood* sequence, or N.K. Jemisin in her *Broken Earth* trilogy.

An idea does not have to be empirically proven to be seized upon by authors. Even as Lovelock's theory has lost scientific credibility, it gained traction as an idea in the environment movement and in popular culture (e.g. Alan Moore's *Swamp Thing*). And, of course, Lovelock's idea of Gaia is a very old one (echoed in mythologies around the world, not just the Hellenic) and was suggested by the author William Golding: literature and nature once again in a symbiotic feedback loop.

Finally, the 'many worlds' theory of quantum physics has spawned countless multiverses within Fantasy. Hinted at in the portals of C.S. Lewis' Narnia prequel, *The Magician's Nephew* (1955), and the 'Eternal Champion' imagined by Michael Moorcock, it has been dramatised most whole-heartedly by Philip Pullman in *His Dark Materials* (1997). This is connected to Entanglement Theory (Wilczek, 2016), which can be summarised as follows: 'An entangled system is defined to be one whose quantum state cannot be factored as a product of states of its local constituents; that is to say, they are not individual particles but are an inseparable whole. In entanglement, one constituent cannot be fully described without considering the other(s). The state of a composite system is always expressible as a sum, or superposition, of products of states of local constituents; it is entangled if this sum cannot be written as a single product term' (2022). One can see how this can be applied to the world-building and plot of a Fantasy novel, and can perhaps be best illustrated by George R.R. Martin's 'Song of Ice and Fire' series (adapted as the HBO series *Game of Thrones*), which conveys this through its portrayal of geocomplexity, the interlinked fates of its protagonists (chiefly the surviving, scattered Stark family members), and the multiple threats of White Walkers, Dothraki nomads, dragons, and feuding Houses.

The geographer Tim Cresswell argues that 'place' is 'a way of seeing, knowing, and understanding the world' (2014: 18), and that 'this way of seeing can seem to be an act of resistance against a rationalization of the world that focuses more on space than place' (2014: 18). Every Fantasy novel has this opportunity—to convey depth, and not just breadth. To aspire to the 'deep mapping' of Cairngorms-based writer Nan Shepherd (2011 [1977])—who spent forty years intimately mapping her home—to chart a place intimately through a long-term engagement with it.

Of course, other ethical vectors can completely change a Fantasy 'landscape' with their 'act[s] of resistance', such as Feminism, NeoMarxism, Deep Ecology, and LGBTQI+ rights—and all are worthy of deeper exploration—but hopefully this selection conveys an idea of how a consideration of any can help provide powerful convection currents for the mantles of the embryonic world being built.

Autocartography and Non-Mapping

As Rebecca Solnit has pointed out, a map is always a political statement, one with implicit (or explicit) power discourses. The map in Fantasy has become a default convention ('. . . no Tour is complete without a Map', Wynne Jones, 2004), but in a new wave of literary Fantastika, its literalising effect has been eschewed for an embedded textual cartography. In the novels, *The Underground Railroad* (Whitehead, 2016); *Lovecraft Country* (Ruff, 2016); *Under the Pendulum Sun* (Ng, 2017); *Who Fears Death* (Okorafor, 2010); *Black Leopard, Red Wolf* (James, 2019); *The Buried Giant* (Ishiguro, 2015); and *The Power* (Alderman, 2016), we see how powerful and evocative Fantasy can be when it eschews the comforting tropes of its genre. Primarily at the literary end of the spectrum (Hilton, 2024), one of the signifiers of this different agenda is the absence of a map.

In this section I wish to explore the implicit self-mapping and mapping of self that these authors articulate—what has been called 'autogeography'. In 1966 the artist Saul Steinberg created a hybrid landscape-map, entitled 'Autogeography'. Avi Steinberg describes it as follows:

> 'Over a plain meanders the river of Steinberg's life, weaving among dozens of place names sized and sited according to the vagaries of fond or bitter memory: Râmnicu Sărat ('a place invented for me to be born in'); Tashkent and Tuba City ('where I bought a hat'), Riverhead and Calcutta, Milan and New York. Here, as so often in a career spanning six decades, Steinberg used a language of his own devising to recreate the world in his own image.'
>
> (2011)

Since Steinberg's creation of the term, autogeography has widened to include other media, and in literature it has come to mean a form of embodied writing, drawing upon the physical actuality and intersectionality of the author. This does not limit the writer to the mimetic, as the examples I will discuss here prove. Such prima materia is ripe for self-manipulation and transformation through the imaginal lens. Virginia Woolf in *Orlando* (1928) imagined a protagonist who altered their gender to thwart misogynistic inheritance laws. In *The Metamorphosis,* Franz Kafka (1915) dramatised the morning when Gregor Samsa woke up transformed into a giant insect. Both texts comment upon the actuality and limits of selfhood via thought experiments in alterity.

By the 19th century, realism had become the default mode of the novel, and despite the anti-mimetic experiments of the Modernist novelist, it still established itself as the dominant mode, with Fantasy ghettoised along with other genre fiction. Yet, the slavish materiality of mimetic fiction and its attendant claim to 'authenticity' is just another, lesser, kind of fantasy—the fallacy of omniscience—as Le Guin epitomises with her usual piercing insight: 'Fake realism is the escapist literature of our time' (1979). Novels that eschew the exhaustive mapping (of the mundane) of Realism are attempting something else. Milan Kundera, in his exposition of his craft, *The Art of the Novel* (1988), critiques the obsessive rendering of detail in mimetic fiction, suggesting that a novel is vehicle for exploring 'not reality but existence'. Instead, he suggests the Coleridgean 'suspension of disbelief' that we are complicit in is not exclusively predicated upon verisimilitude. The reader is more sophisticated than that, he implies, and even with the slightest of clues can invest characters and a situation with meaning and emotion:

> 'Understand me, I don't mean to scorn the reader and his desire, as naïve as it is legitimate, to be carried away by the novel's imaginary world and to confuse it occasionally with reality. But I don't see that the technique of psychological realism is indispensable for that.'
>
> (Kundera, cited by Turchi, 2004: 172)

Instead, Kundera and other 'postrealist' (to use Turchi's term for postmodern) authors such as Borges, Nabokov, and Calvino offer narratives that examine themselves, as well as the readers' role in their 'creation': 'Postrealistic work simply ask us to imagine other ways of imagining. Some of it draws the reader's attention to the artificial nature of its construction. . . . At the same time, explicitly or implicitly, postrealism takes as its subject our ability to imagine ourselves in a world created entirely by the placement of ink on a page' (Turchi, 2004: 172).

The writer could be said to tattoo the page with ink—with the marks of their handwriting, typewriter, or word processor. Simultaneously, a ghost tattooing is taking place in the reader's mind during the act of reading: marks are turned into words into meaning—and these join together to create narrative, one brought alive through the 'mind's eye'. The visualisation of the author becomes the visualisation of the reader: the page (or screen) is just the intermediary via a kind of psychometric act—textual telepathy. In this sense all fiction involves an element of 'fantasy': our 'image-making faculty' (Concise Oxford Dictionary)—in its creation and consumption. Fantasy writers just take this to the extreme, but conceptualising and rendering new or alternate worlds and ways of being. The literalising effect of a visual adaptation can colonise the imagination, and foreclose on the very Fantasy it is promoting—the individualised Gandalf created by every reader of *The Hobbit* and *The Lord of the Rings* becomes one concretised by a particular actor in a particular version, however well done. Hence, the nervousness of fans when a new adaptation or iteration of the 'franchise' is announced, for example, the posting of the first official image from the Amazon series based upon The Second Age of Middle-earth elicited a spectrum of reactions—from excitement to prayers that they 'get it right' (Twitter, 2nd August 2021).

The Fantasy occurs not in the presence of a map, or in the presence of magical beings, places, and incidents, but in our *imagining* of them—and, critically, in the intentional 'gaps' left by the author, in plots, in behaviour, in dialogue.

The most evocative fantasy is often created by not describing what is there, but by alluding to what is *not* there. This 'telling absence' is what Mark Fisher identifies as the difference between the 'weird' and the 'eerie'. In the former, something is present that should not be there; in the latter, something is absent that should be there: 'The weird, by contrast, is notable for the way in which it opens up an *egress* between this world and others' (2016: 19). Of course, at its worst, this can lead to H.P. Lovecraft's accidentally comical, and much mocked phrase 'the-thing-that-cannot-be-describe d', but the use of spaciousness in writing can increase, not decrease the Fantasy—something would-be writers should consider before rushing to populate their world with exhaustive world-building. Ultimately, there is room for a spectrum of approaches—the putter-inner and taker-outer and everything in-between—and every writer needs to work out the optimum balance. It is a delicate tight-rope walk, as Turchi suggests:

> 'Readers get one kind of pleasure from a complete shape and another kind of pleasure from perceiving a shape only partially represented—from playing a role in bringing the work to completion. As writers we need to determine how much of the shape to supply, how clear to make it—and to understand how much work we're asking the reader to do. If we provide too little, we fail to communicate. If we provide too much, there is no room for the imagination. But when we get it right, we're the best of guides, leading the way to a place that allows for the reader's discovery'
>
> (2004: 204)

This notion has been formulated into what has become known as 'Hemingway's Iceberg', which, taken to an extreme, can create a cold, flinty 'macho' form of prose, exactly like Hemingway's hard realism, but as a rule of thumb, for the creation of subtext, it has wider usage. With this caveat in place, Hemingway's approach is worth citing in full here.

> 'I always try to write on the principle of the iceberg. There is seven-eighths of it underwater for every part that shows . . .
>
> If a writer of prose knows enough about what he is writing about he may omit things that he knows and the reader, if the writer is writing truly enough, will have a feeling of those things as strongly as though the writer had stated them . . .
>
> [But] if a writer omits something because he does not know it then there is a hole in the story.'

Such 'holes' in stories lead to unsatisfying or frustrating reads. And so telling absences or negative spaces must be deployed cautiously and with what could be called 'depth-knowledge' of their narrative, storyworld, and the inner worlds of their characters. Like a conscientious actor who has done a lot of preparation before their performance to find a way into their character, to flesh them out, to fully inhabit them, to know them inside out (Stanislavski, 2022)—deploying this kind of 'method-writing', writers with that sense of depth-knowledge can use the lightest of touches, but can evoke an undertow of associations, a polysemous resonance, and thus achieve a graceful complexity.

Post-colonial Counter-mapping and the Regendering of Space

I now wanted to consider post-colonial and late feminist strategies outside of Fantasy that can offer innovative approaches to the 'mapping' of imaginary worlds.

'Counter-mapping' is a term coined by Nancy Peluso in her 1995 article, 'Whose Woods Are These?', in which she develops the notion in the context of an NGO-led environmental campaign to protect the forest territories of Kalimantan, Indonesia. Within a contested space (Reyes, 2016), alternative geo-spatial narratives are offered, often from a marginalised indigenous perspective, and often eschewing conventional forms of mapping, instead deploying song, story, myth, ritual, dance, and art. The term has equivalents and precedents in bioregional mapping, parish-mapping (such as Common Ground's ideation of 'local distinctiveness'). Eades and Zheng consider it a form of 'assemblage':

> Moving across time and space, counter-mapping assemblages form dynamic connections among trans-local knowledge systems as well as

> older and younger generations, reconnecting the lifeworlds of modernized indigenous groups and their history, tradition and natural territories, thereby reconfiguring their identities.
>
> (2014: 92)

Robert Macfarlane, discussing the indigenous 'counter-mapping' of indigenous people, such as the Inuit in the Canadia Arctic, or Koyukon people of northwest interior Alaska, who navigate via a spoken map of story, observes, 'In such maps, human memory and natural form recoil into one another' (cited in Turchi, 2004: 101).

An inspiring post-colonial counter-mapping initiative is the Zuni Map Art Project, based in New Mexico. This is led by Jim Enote, a Zuni farmer and director of the A:shiwi A:wan Museum and Heritage Center. In a film-essay by Adam Loften and Emmanuel Vaughan-Lee ('Counter-Mapping', Vaughan-Lee and Loften, 2019), Jim introduces us to his project as we follow him on his daily rounds to his farmland and centre, all within his ancestral homelands. His deep knowledge and respect for the land are evident, as he converses with his crops, 'checking in' on them, and pays homage to the elemental spirits. He observes, 'Modern maps don't have a memory' and 'We limit ourselves if we only think of maps as two dimensional.' To counter this, Jim explains how he has 'patterned languages that help [him] to remember how [to] get from one place to another' ('Counter-Mapping', Vaughan-Lee and Loften, 2019). This lexical cartography echoes the Aboriginal songlines, and like their Dreamtime art tradition, Jim has commissioned Zuni artists to paint their own memory maps of the landscape. These are maps that convey a subjective sublime—depicting genius loci, important rituals, significant historic events, and culturally important figures within their community and tradition. Loften and Vaughan-Lee describe how '[t]he Zuni maps are an effort to orient the Zuni people, not just to their place within the landscape, but to their identity, history, and culture. The maps contain a powerful message: you have a place here, we have long travel[l]ed here, here is why this place is important. Through colo[u]r, relationship, and story, the maps provide directions on how to return home' ('Counter-Mapping', Vaughan-Lee and Loften, 2019). To date, Jim has commissioned thirty of these maps, and more are being created and exhibited in his centre. Thereby the map-paintings become a resource for teaching new generations about Zuni tradition, while simultaneously providing a platform for their contemporary indigenous art scene (and vital income). The initiative is empowering, educational, and practical. Art can be an extremely effective tool in community development—a celebration of local distinctiveness and cultural identity. It can also be used to critique hegemonic discourse and endemic bias, and facilitate essential framing of collective space. At the forefront of this is American writer Rebecca Solnit, and her work in regendering the city.

In much of her work Solnit has questioned the heterodoxies of culture through her erudite enquiries into walking, art, literature, disaster capitalism, the Climate Crisis, and gender, although she admits in an interview with Emma Watson (2019) to have 'held back' from adopting an overtly feminist critique in her early oeuvre—a reservation she no longer has. In her talk for the Creative Time Summit of 2013, 'Art, Place, & Dislocation in the 21st Century City', Solnit offered, 'A Thousand Stories in the Naked City' (2013) which looked at how the naming of a city—its streets, buildings, public spaces—can imply certain power discourses, and how therefore a regendering of the city can be radical act that can shift consciousness. She discusses her project (and collaboration with geographer Joshua Jelly-Shapiro) to rename the stations on the New York Metro with women (Cascone, 2016)—feminist icons, public figures, role models—and how this changed how her female students from Columbia felt when exploring the urban landscape. In her interview with Watson, Solnit playfully suggested that Watson did the same with the London Underground, a project which the young actor has now instigated with the author Renni Eddo-Lodge, who took to Instagram to ask, 'If your local London Underground station was named after a woman, who would it be?' (Devaney, 2021). Such initiatives, especially when devised in such an egalitarian way, enable citizens to redefine (remap) the landscape in which they live. This conveys ownership, and a rebalancing of power dynamics. The ongoing debate about the, often problematic, figures represented by public statuary—a debate that is perhaps understandably emotive, as witnessed by the debouching of slave-trader Edward Colston, in Bristol harbour by Black Lives Matter protesters in the summer of 2020 (BBC, 2020)—shows how vital it is to consider the way we name, and frame our landscapes we live and work in (Olusoga, 2016; Fowler, 2019).

In sharing these examples—the Zuni Map Art Project; Solnit and Jelly-Shapiro's New York City tube map—the intention has been to illustrate how such creative-critical approaches can help 'hack' the tacit hegemonic discourse of all kinds of maps. With important real-world applications, such approaches can also be applied to the mapping of imaginary spaces (as in Secondary World novels). And it could also be a way of directly critiquing existing, iconic ones, for example, what would a counter-map of Middle-Earth look like from the perspective of an orc, or an Easterling? What would a regendered Arthurian landscape look like?

Some texts have attempted this with results that are both innovative and engaging, for example, John Gardner's *Grendel* (1971), which reimagines the Anglo-Saxon poem of *Beowulf* (a key influence on Tolkien's legendarium) from the point of view of its main antagonist; Marion Zimmer Bradley's *The Mists of Avalon* (1983), a feminist revisionist version of the Arthuriad that retells the story from Morgaine/Morgan le Fay's perspective, to name but two. Now, there is a whole slew of novels retelling well-known myths and legends from fictitious marginalised perspectives, for example, Manda Scott's *Boudica* (2003); Madeleine Miller's *Circe* (2019). This mirrors the Hollywood 'spin-off' movie, a business model designed to milk a franchise for all its worth. Although some novels, comic books, and movies are more daringly conceptualised and better written than others, it has become formulaic schtick to power up a minor, or disempowered character and give them their own 'tragic backstory' (Smith Spark, 2025), whether it is a supporting character (Van Helsing, Igor, Hansel and Gretel, Red Riding Hood, Ophelia), villain/ess, or 'sidekick'. When done well, as in Tom Stoppard's play based upon two 'flat' characters from Shakespeare's *Hamlet, Rosencrantz and Guildenstern are Dead* (1966), *The World's Wife* by Carol Anne Duffy (1999), or *The Penolopiad* by Margaret Atwood (2005), the approach can result in high art, or at work that is entertainingly iconoclastic, for example, *American Gods* by Neil Gaiman (2001); *The Gospel of Loki* (2014) and *The Testament of Loki* (2018) by Joanne Harris; *Thor: Ragnarok* (dir: Taika Waititi, 2017); *Loki* (Marvel/Disney, 2021), which all riff on Norse mythology in a neo-medievalist, postmodern manner.

Counter-mapping, regendering, and other forms of 'reframing' (Stibbe, 2021: 41) are effective methods for addressing the ethical imbalances of a mapped territory.

Now, let us consider the pervasive notion in Fantastika of 'world-building'—another shibboleth that needs to be challenged.

Against World-Building

An argument *against* the very notion of world-*building* (the linguistic framing of such suggestive of an Imperialist, colonial project) has been articulated briefly by Hergenrader (2019: 5–6), but perhaps a more critically substantive critical position can be found in the work of post-structuralism of Deleuze and Guattari, chiefly in their concept of what they call 'deterritorialisation' (notably in *A Thousand Plateaus,* 1988). In one of the 'plateaus', as they call the sections of their book, 'On the Refrain', they explore 'nature', art, and the nature of art, in relation to birdsong and avian behaviour. Discussing how a bird's song can change from one with a territorial dimension to a different kind when seeking a mate, they note how the outpouring transforms from boundary-marking to door-opening. This leads to their advocacy of 'a deterritorialized refrain as the end of music, release it in the Cosmos', because 'that is more important than building a new system' (Deleuze and Guattari, 1988: 350). Their mechanism for achieving this is what they call an assemblage, an ephemeral grouping of disparate elements, echoing Hakim Bey's 'Temporary Autonomous Zone' (1991). Like Bey's anarchic pirate utopia, an assemblage 'holds heterogeneities together without their ceasing to be heterogeneous' (1988: 329). This offers a more fluid, wide-ranging model than Foucault's 'heterotopia' (1997), which is site-specific, and perhaps is closer to the Zapatista slogan, 'Por un mundo donde quepan muchos mundos' ('For a world in which there is room for many worlds') (2001).

Deleuze and Guattari use the analogy of Chess and Go to delineate the difference between what they call 'striated' and 'smooth' space. In the former, the board is rigidly regulated grid, with the key aim of the game being the capturing and controlling of space. Whileas,

> [i]n Go, it is a question of arraying oneself in an open space, of holding space, of maintaining the possibility of springing up at any point: the movement is not from one point to another but becomes perpetual, without aim or destination, without departure or arrival. The 'smooth' space of Go, as against the 'striated' space of chess.
>
> (1988: 353)

This could suggest two significantly different approaches to the writing of Fantasy, and the rendering of setting—one is highly planned and controlled, the other is exploratory and discursive. However, such a methodology would not suit everyone. Their 'smooth space' may seem to some, a bridge too far.

However, the latter approach has been framed (in an exegesis on his own writing practice) by Harry Whitehead as 'nomadic emergence' (2013: 101), and indeed by Deleuze and Guattari themselves in their *nomadology* as 'nomad space':

> The nomad, nomad space, is localized and not delimited. What is both limited and limiting is striated space, the *relative global*. . . .Even when the nomad sustains its effects, he does not belong to this relative global, where one passes from one point to another, from one region to another. Rather he is in a *local absolute* that is manifested locally, and engendered in a series of local operations of varying orientations: desert, steppe, ice, sea.
>
> (1988: 382)

This concept of the 'local absolute' is echoed in LeFebvre's ideation of 'absolute space' (as part of a trialectic of absolute, abstract, and differential space analogous to past, present, and future). In absolute space, a kind of prelapsarian idyll is co-opted and exploited by outside forces. LeFebvre describes this in what could be a precis of many a Fantasy novel:

> The cradle of absolute space—its origin, if we are to use that term—is a fragment of agro-pastoral space, a set of places named and exploited by peasants, or semi-nomadic pastoralists. A moment comes, when, through the actions of masters or conquerors, a part of this space is assigned a new role, and henceforward appears as transcendent, as sacred (i.e. inhabited by divine forces), as magical and cosmic.
>
> (LeFebvre, 1991: 234)

Stephen Hardy provides a gloss on this, describing the process as 'a political appropriation of a certain space, which is then used to form the dominant centre of a new, more organised kind of social space' (Hardy, 2000). The 'smooth' becomes 'striated'. The plot of many a Fantasy novel is contingent upon the restoration of this original state—a re-enchantment, one that would scour the Shire or heal the Wasteland. It is the Campbellian 'return with the elixir' (2000 [1948]), although like Gilgamesh himself, many such hopes have expired on the difficult 'road back', and in late Fantasy (e.g. Grimdark), the re-enchantment is often stalled, or abandoned altogether. The world is broken, and like the importunate Humpty Dumpty, nothing will put it back together again.

Although many still find comfort within the formula of the genre-safe, deterritorialised Fantasy (such as seen in the work of Okorafor, James, Ishiguro, Harrison, Alderman, Ng, Joyce, Kehlmann, VanderMeer, Clarke, and Gaiman) offers exciting possibilities. Resisting the fallacy of omniscience, and actively critiquing the hegemonic discourses of classic Fantasy, it offers a defamiliarised territory in an over-saturated market—one in which the striated demarcations of genre convention are erased and the restored tabula rasa of the Fantastic once again explodes with possibility.

The View Back

An exploration of the way setting or place in Fantasy is constructed, depicted, and mapped is, by definition, never going to be comprehensive—such expansive territory is perhaps the ultimate fuzzy set, and is being continually added to. In such an instance 'completeness' is as specious as any Grand Map. It is an unending quest. As Donovan once sang, 'First there is a mountain, then there is no mountain, then there is.' The territory exists before the map, but is recreated by the map in a highly selective and contested fashion, facilitating both discovery and elision. And yet it is hoped this brief survey will provide some lembas for the journey.

We started this chapter by asking three questions. So, what has our enquiry yielded? Not *every* Fantasy begins with a literal map, although many are inspired indirectly by the maps encountered in formative reading (Robert Macfarlane, Cressida Cowell, Frances Hardinge to name a few). Certainly, a desire to 'explore, to know, to record, and to share' is, if not universal, certainly very common amongst writers. All humans are self-reflexive sentient beings—who dream and desire (and by doing so map their emotions onto 'that what is not', onto *terra nondum cognita:* Lands Not Yet Known), but writers are adept at imagining with greater exactitude and finesse, or at least articulating the inchoate worlds of the consciousness better than most. They are the pioneers, the trailblazers of imaginary realms, which then become colonised by the readers who follow. Some of these realms become mapped by subsequent waves of critics, scholars, and creatives. I also asked, 'Do maps foreclose on Fantasy, colonising our imagination before we've even started reading?' The highly selective, hierarchical finitude of a map both focalises our attention on what is included, and begs the question of what is left out and why. As Pratchett has quipped, 'Anyway, what is a fantasy map but a space beyond which There Be Dragons', acknowledging the cartographical cliché first established by the Hunt-Lenox globe of 1510—one of the only *historical* maps to actual bear the legend, *H[i]c Sunt Dracones*—a kind of geographical marginalia acting as placeholder for lack of available knowledge: the high-water mark of hubristic omniscience. Yet, some would argue the opposite, emphasising the way maps inspire imaginative engagement and creative response, such as Cressida Cowell: 'A map is . . . a story starter, an idea generator. Drawing a map is a magnificent way of communicating with your unconscious' (cited in Lewis-Jones, 2018: 85). The novelist David Mitchell also emphasises their usefulness in the creative process: 'This is why mapmaking and 'stage-sketching' can be necessary aspects of writing' (Lewis-Jones, 2018: 120). Maps of imaginary places, in particular, seem to elicit an often passionate imaginative engagement with the hypothetical territory. Mitchell again: 'Fictitious maps give form to a thing—the imagination—that has no form. They are mysteries and answers to those mysteries' (Lewis-Jones, 2018: 124). For children's writer Piers Torday, maps, even highly fictionalised ones, were 'essential certificates of authenticity to accompany journeys into fantasy landscapes' (Lewis-Jones, 2018: 132). Abi Elphinstone also advocates the importance of speculative cartography: 'I begin every story I write by drawing a map because it is only when my characters start moving from place to place that a plot unfolds' (Lewis-Jones, 2018: 147). Joanne Harris echoes this notion, exploring the etymology of the word 'plot': 'It is no accident that "plot" can mean at the same time the arc of a story, or a chart showing the course of a ship, or the tracing of a map. These things are interconnected. The idea of movement, of laying out, of following a set path—all these things are part of the language of human exploration as well as that of narrative' (Lewis-Jones, 2018: 110–111). So, for many writers, different forms of 'mapping' seem intrinsic to the act of creation, and for many readers, the Fantasy map is both the portal and the passport, the 'certificate of authenticity', albeit an authenticity that is highly fictionalised.

Finally, I asked, 'isn't any attempt to "map" this terrain theoretically just another kind of mapping—one that some writers of Fantasy either dislike or actively resist?' Yes, I believe it is another kind of mapping, but as with the Fantasy map itself, it does not shut down the imaginarium around a Fantasy text, but in some ways enlarges or enriches it—as I hope this brief survey has here. Like travellers sharing differing accounts of places visited with their fellow wayfarers, a plurality of narratives does not cancel out the original 'road' of the ur-text. The territory still exists, and each new arrival will develop their own map of it—and each of these personalised maps overlaps with what has gone before and what is yet to come, thereby joining 'some larger way/Where many paths and errands meet', as Bilbo Baggins sings, as he leaves The Shire for Rivendell:

> The Road goes ever on and on,
> Down from the door where it began.
> Now far ahead the Road has gone,
> And I must follow, if I can,
> Pursuing it with eager feet,
> Until it joins some larger way
> Where many paths and errands meet.
> And whither then? I cannot say.
> (1991 [1954]: 44)

In this chapter we have looked at the following 'landmarks': every location and locale in a Fantasy narrative suggests vectors of ethicality; the act of mapping can imply all kinds of implicit and explicit power discourses; placiality changes our perception (and depiction) of place; critical theory offers its own 'maps' to frame the territory; BIPOC authors are deconstructing, resisting, remapping the territory; and the worlds we imagine come with a creative responsibility.

Increasingly, writers of colour are redrawing the (Fantasy) map by drawing upon a wide array of cultures beyond the Global North, for example, S.A. Chakraborty's *Daevabad Trilogy* (2017–2020); Premee Mohamed's *Beneath the Rising* trilogy (2020–22); Fonda Lee's *The Green Bone Saga* (2017–2021); and Marlon James' planned *Dark Star Trilogy* (which began with *Red Leopard, Black Wolf* (2019) to name a few. Such writers are breaking trail and deconstructing hegemonic striations implicit in some Secondary Worlds. Each is an act of empowerment for author and reader, and offer critiques of both nondiegetic dominant discourses, and the implicit power discourses of both the novel and the wider genre. By their resistance to the visual 'closure' of the map, with its colonialist fallacy of omniscience, the novels exhort a self-defining approach; one that also empowers the reader to actively visualise its unillustrated pathways: a cognitive engagement that offers the profoundest affect of Fantasy: imagining the unknown.

Setting is a major asset and allure of Fantasy. It provides as big a pull as the characters and stories themselves. The worlds Fantasy authors fashion, and the way they choose to represent them, affects the whole endeavour, in the way that climate (rather than incidental weather) has a significant impact on the real-world story of the inhabitants (human, and more-than-human) of planet Earth. Whether we choose to map them or not, we cannot ignore their impact.

7 Fantasy Sites

Fantasy Texts and Their Real-World Impact

Real-world locations have been inspiring the creation of Fantasy texts for a long time, and these texts in turn (in whatever form they are transmitted: print, TV series, film, computer games, TTRPGs, graphic novels) can then go on to have significant real-world impact. This chapter considers this phenomenon—what could be called the Fantasy Feedback Loop—and discusses a dozen real-world examples based primarily upon field research.

A writer searching for inspiration, or just feeling restless—maybe suffering from insomnia, wanderlust, or *sehnsucht* (the German noun for longing; desire; yearning; craving)—goes for a walk. Perhaps they are lucky enough to live by an Area of Outstanding Natural Beauty, a Site of Special Scientific Interest, a National Trust property, or a National Park—but more often than not, it is somewhere more modest: a local municipal park, a tow path winding its way alongside a greasy canal between gas work towers, the scrubby edgelands of a brownfield site, or some post-industrial ruin (Farley and Roberts, 2012). Not all of us can 'wander . . . lonely as a cloud, as that serious walker-poet, William Wordsworth was wont to do, lacking the white male privilege; the access to the countryside or safe green spaces: free of potential assault, harassment, prejudice, xenophobia' (Fowler, 2019; Sethi, 2021); the lack of 'high-end technical gear', or 'enough disposable time to wander without any cause other than "wanting to"' (Lanham, 2022). Yet as the parallel traditions of the flâneur/flâneuse (Solnit, 2001; Elkin, 2016; Overall, 2021) and psychogeographer (Coverley, 2018) suggest, inspiration can be found in the derives and soundings of the urban writer—and any walk outside involves nature of some kind, whether it is the weather, the weeds in the pavement, gutter or verges, urban wildlife, designated green spaces, gardens glimpsed through railings or on window-boxes, or even 'just' the air quality, temperature, humidity, or quality of light and shadow. There are numerous examples of writers finding inspiration in the urban—and in urban nature in particular: William Blake beholding angels in a tree in Lambeth; J.M. Barrie conjuring Peter Pan in Kensington Gardens; Arthur Machen embarking on his

DOI: 10.4324/9781003463665-8

'London Adventure'; or WB Yeats being reminded of 'The Lake Isle of Innisfree' by a civic fountain in a London street—to name just a handful of examples from one capitol city. But for the purposes of this chapter I will focus on wider-ranging classic examples from what is generally designated the 'countryside', or 'landscape' (both relatively recent cultural constructions) as their impact can be more easily discerned—in terms of both influence and specific visitors (unlike an urban space, which will have many visitors and passers-by with a multitude of reasons for being in that area).

For the purposes of this chapter, I classify 'Fantasy sites' as sites associated with a particular Fantasy text—novel, TV drama or series, film, computer game, or graphic novel—in terms of initial inspiration, adaptation, or other association (such as source material for a computer game setting). I acknowledge the possible confirmation bias of my chosen data set. Similar surveys of sites could be conducted in many locations across the world, and would be equally interesting, although the long-established nature of some classic Fantasy texts means there is a more discernible 'pull' and impact than more recent or less well-known works. However, it would certainly be interesting to see a survey of works that eschew the Eurocentric bias of many of these.

My methodology in investigating these sites draws upon my experiential approach to writing in general, which I have developed over a number of years and have discussed at length elsewhere (Manwaring, 2020a, 2020b, 2020c). In brief, I believe in the value of the visceral, immersive, embodied experiences; the discovery of telling details in situ; the savouring of the affect of a place, qualia-capture via notebooks, audio notes, and other documentation; and research of the palimpsest of narratives associated with a site combined with the phenomenological actuality of being there for real—in the flesh, and not merely virtually. This may be digital dualism (although it is a problematic concept in itself, one that makes many First World assumptions about access to devices and Wi-Fi), but the appeal of such sites is that there is no signal, and, if one chooses to disable 'location' on one's mobile phone, blissfully invisible to the Sauron-glare of the communications satellite, at least briefly.

During the summer of 2022, with the help of an RKE (Research Knowledge Exchange) Fellowship from Arts University Bournemouth, I was able to undertake a series of field trips throughout England, Wales, the Republic of Ireland, and Northern Ireland, to visit a range of Fantasy sites. I drafted a survey form to record essential details and impressions in a standardised form; I took over 1,500 photographs at the dozen or so sites I visited; and I made field notes in my notebook too (including the occasional poem, or short story idea). Practically, I wasn't able to conduct extensive interviews or exhaustive research into the socio-economic impact of the Fantasy texts in these sites, having only a two-week window, so I focused primarily on the 'affect'—something which cannot be gleaned vicariously. Others may choose to do more comprehensive sociological research and with greater authority than I could offer as an Arts, rather than a Humanities, scholar. I am merely breaking the trail here.

My field research framed and formalised what I have been intuiting for years: that certain sites have certain special qualities conducive to inspiration—and often these 'special qualities' are a combination of prior reading (myths, legends, folk tales, novels, graphic novels) and viewing (documentaries; TV series; films) and the genius loci—whatever we perceive that to be. Writers, artists, and musicians have been compelled to evoke that for centuries, as Susan Owen convincingly charts (2020).

My survey form collected the following: the date; primary location; weather conditions; associated texts; type of site: primary source (places of inspiration and composition); secondary source (places associated with film/TV locations); tertiary source (visitor attractions that artificially recreate the aesthetic of Fantasy texts). Also, the manner in which I documented the 'affect': photography; audio notes; ambience; field notes; sketching. Then I ask about signage—Is there any signage highlighting its status? Interpretation panels? Pamphlet, or app? Visitors? Any visitors? Are they aware of the associations? Eco-impact (negative)—Is there litter? Other signs of environmental degradation, for example, erosion of paths; damage to buildings? Eco-impact (positive)—Any conservation efforts? Protected status? Creative/critical responses—Does it inspire any ideas? Fiction? Poetry? Reflections—Does the visit change your perception/appreciation of the text/author/adaptation? And finally, Miscellaneous—is there anything else worth mentioning?

With that in mind, now let us consider the example Fantasy sites.

Avebury

We will begin our brief survey of a selection of British and Irish Fantasy sites with one of the oldest—Avebury, a significant Neolithic henge monument and stone circle(s). The Great Circle is one of the largest in Europe and the largest in the United Kingdom. Avebury and its environs, along with Stonehenge and its landscape, together form a UNESCO World Heritage Site. Although the latter is more widely known (a modern 'Wonder of the World' that seems to be on most people's bucket-list) and iconic with its unique trilithons, and certainly a much-used landmark in popular culture as the 2022 British Museum exhibition showed Avebury is more apposite to our needs here with its association with a 'cult' Fantasy text. The central monument actually comprises three—each built in a separate phase. Archaeologist Aubrey Burl suggests dates of 3000 BCE for the central cove (comprising a pair of standing stones); 2900 BCE for the inner stone circle (a medium-sized circle, now mainly comprising concrete markers); and 2600 BCE for the outer circle (1088 ft across; originally comprising 98 sarsens from the nearby glacial erratics of Fyfield Down) and henge, and around 2400 BCE for the avenues (Burl, 2002: 154). Beyond this vast hub (which contains part of the village of Avebury itself including The Red Lion pub), there is a wider temple landscape with a network of sites: Windmill Hill; The Avenue; Silbury Hill; West Kennet long barrow; The Sanctuary; Overton Hill round barrow cemetery; The Ridgeway; and The Devil's Den to name the most significant. How these all relate is a matter of debate (or wild speculation), as there are sometimes significant differences in construction dates ranging from the early Neolithic to the Bronze Age, for example, at 3675 BCE, Windmill Hill predates the earliest phase of Avebury by 675 years. Nevertheless, they collectively create an uncanny landscape—one that has fascinated writers and artists for centuries.

Since the Wiltshire-born antiquarian and Fellow of the Royal Society, John Aubrey, 'discovered' Avebury in the 17th century while out on a hunting trip (or at least brought it to the attention of his erudite circle and Royalty, piquing the interest of no less than Charles II, who commanded Aubrey to show him the area), it has drawn lively enquiring minds like a magnet. Aubrey's 18th-century equivalent, William Stukeley, undertook the first thorough survey of the site, executing priceless sketches showing the stones as they appeared at the time (including some that have been destroyed altogether, such as The Sanctuary).

In the 20th century, Avebury drew to it post-war artists in search of re-enchantment. War artist Paul Nash (1889–1946), raw from depicting the denatured hellscapes of the First World War (Walton, 2022), found fresh inspiration in the strange megalithic structures standing stark against the clean line of the Wiltshire downs (echoing the shell-blasted stumps of the Somme)—rendering them with an English Surrealism. Graham Ovenden (born 1943–), member of the Brotherhood of English Ruralists, continued this conversation—evoking an 'English Eerie', one that captured the cthonic numinous energies of Avebury perfectly, but with the hindsight of his 2013 conviction for indecency and indecent assault against minors could be seen as a foreshadowing of his own troubled mind. The sense of a disquieting immanence was also captured by another English artist, Derek Jarman, who recorded his 'A Journey to Avebury' (1971) on Super 8mm film, which he later tinted and edited into a short film (10 min., 45 sec.). In this, the surrounding landscape is given equal significance to the stones themselves, and there is a sense of a kind of intuitive Pagan pilgrimage—perhaps one undertaken by early visitors along the Ridgeway, the 'oldest road in Europe'. Poet Adam Thorpe, who attended the nearby Marlborough College as a boy, created his own propitiatory offering in print to the genius loci of Avebury, *On Silbury Hill* (2014)—a memoir of place and his formative years growing up within its radius. Many other artists and writers have been inspired by Avebury, but I would like to focus on two main associated Fantasy texts: *Children of the Stones* (1977); *and Guardians of the Tall Stones* (1977–1986).

Children of the Stones has become a 'cult classic'. A children's TV series created by HTV in 1976 and broadcast on British television during 1977. It was directed by Peter Graham Scott and co-written by Jeremy Burnham and Trevor Ray. The story follows scientist Adam Brake and his son Matthew, who arrive in the sleepy English village of Milbury—loosely based upon Avebury, Wiltshire—to find it under the grip of weird psychic powers unleashed by the sinister village squire, Hendrick, and whose power they struggle to break. With its unsettling blend of Folk Horror and SF, it evokes elements of *The Midwich Cuckoos* (1957), *Quatermass* (1979), *Doctor Who* (1963–1989; 2005–), and *The Wicker Man* (1973). It has been cited as one of the scariest children's dramas ever made, and was only broadcast once in the UK, although it was released as a film in the US and Canada on the Nickelodeon channel. In 2020 the BBC released a new audio drama adaptation, starring Reece Shearsmith—fully exploiting the audial possibilities of the new form with the son of Brake making his own podcast about village life. The saying of the conspiratorial villagers, 'Happy Day!', has become a metonym for both the show, and Folk Horror in general. Many fans of the series visited Avebury, hoping to imbibe some of its uncanny atmosphere.

The *Guardians of the Tall Stones* is a Fantasy trilogy written by South-African born writer, Moyra Caldecott (1927–2015). It comprises *The Tall Stones* (1977b); *Temple of the Sun* (1977a); and *Shadow on the Stones* (1978). Set in Bronze Age Britain, it is a bildungsroman about 14-year-old Kyra, who has dormant psychic abilities. When the villainous priest, Wardyke, overthrows the beloved elder, Maal, and threatens them all, Kyra must accept that her emerging psychic powers are her people's only hope. With guidance from Maal, and help from Karne and her best friend, Fern, Kyra enters the forbidden circle of stones to call upon unseen, mystical powers to banish Wardyke 'once and for all'. Although lesser known than the Avebury/Milbury-based TV series, Caldecott's trilogy gained its own modest fan base and was one of the pioneer texts of revisionist Neoagan narratives which gained traction during the 'New Age' publishing phenomenon of the late 80s and 90s and drew upon popular 'sacred' sites such as Glastonbury, Arthuriana, Celticity, or loose reconstructions of prehistoric societies. The fact that both of these texts—*Children of the Stones* and *Guardians of the Tall Stones*—appeared initially in the same year, and that *Quatermass: The Final Sequel* (1979) connected a cosmic horror scenario to 'Ringstone Round', a Stonehenge surrogate; and Arthur C. Clarke's *Mysterious World*, a 15-part TV series from Yorkshire Television first broadcast in 1980, featured Stonehenge and other enigmatic sites such as the Nazca—lines prominently in association with a wide-range of popular theories, helped to generate an aura of mystique about such places: one that was particularly (1973) appealing to the Counter Culture. The Stonehenge Free Festival had been what Hakim Bey defined as a 'Temporary Autonomous Zone' (Bey, 1991) since 1974, and Stonehenge and Avebury had become sites of neopagan pilgrimage since the 60s (as memorably evoked in the diary by Stewart Harding and his fellow walkers). Such sites became focal points for the marginalised, disenfranchised, and other hegemonic refuseniks, and for forms of creative resistance to the often antagonistic government (ITN, 1985).

I had been visiting Avebury since the late 80s, and in late 2019, I moved within 5 miles of Avebury—wanting to live in the landscape that had many good memories. Revisiting it for the field research, I had to defamiliarise myself with it and try to see it with fresh eyes—but this was virtually impossible. It is such a storied landscape. I was aware of its rich palimpsest of narratives, but it was certainly Caldecott's trilogy which focalised my appreciation of it, preceding (for me) *Children of the Stones*, which I didn't watch until relatively recently, shortly after moving to the area. The HTV series definitely bestowed upon the stone circle a certain 'cultic' quality (folk horror affect), despite it being ostensibly a benign, peaceful place. I had been raised on a diet of programmes like Arthur C Clarke's *Mysterious World* (1978) and *Quatermass: The Final Sequel* (1979); and classic reading such Alfred Watkin's *The Old Straight Track* (1925); Michael Dames' *The Avebury Cycle* (1977); and *The Silbury Treasure* (1976), and more recently Adam Thorpe's *On Silbury Hill* (2014). And my appreciation of the works of Paul Nash and Derek Jarman gave Avebury a certain aura of Fine Art kudos. The site is now co-managed by English Heritage and the National Trust. It is very popular in the summer. This results in erosion on some of the grassy areas around the stones and on the surrounding 'henges' (raised banks and ditches), and so erosion-control is often in place. The key sites have corresponding interpretation panels, and the connecting trails are well signposted and generally well maintained. There is occasional litter, but most visitors are respectful, with only the odd incident of vandalism. Considering that at one point the systematic destruction of the stones was very much the societal norm (during the 17th and early 18th centuries), this shows a significant shift of perception. The various Ancient Monuments Acts protecting such sites (starting in 1882, as instigated by John Lubbock, 1st Baron Avebury himself; and updated in 1900, 1913, 1931, 1979 to date) certainly helped dissuade members of the public from damaging them—although this still happens. But it is their depiction in popular culture which has both increased their appeal and their cultural significance—a double-edged sword. Avebury does not receive anywhere near the number of visitors of Stonehenge, and unlike Stonehenge is free to enter (except for parking), and yet the revenue created by Stonehenge helps to conserve its sister sites across England. Visitors are drawn to Avebury for a wide-range of reasons—a pleasant place for a picnic, to walk the dog, to take photographs, to attend a Pagan ceremony, to watch the sunrise at the summer solstice, to dowse for 'leylines', to paint, to write, to visit the pub, gift shops, and National Trust tearooms, a village event, and so on. It is a broad church—but the fact that it is perceived as being 'special', and not just a quirky bend in the road—has no doubt helped to preserve it. Thanks to the efforts of Alexander Keiller in particular (the marmalade magnate, who bought up much of the village in the early 20th century and whose museum now stands in the National Trust 'village'), the site has been preserved for future generations. It is a site where the magical and mundane co-exist—sarsens, ice-cream, and crystal shops—and this is a pattern we will see at other sites on our field trips.

Puzzlewood and Lydney

Puzzlewood is a 14-acre area of ancient forest near Coleford in the Forest of Dean, Gloucestershire. The collapse of an underground river system created a natural labyrinth of twisting mini-gorges nestled between tangled roots and moss-lined rocks. These were exploited by the Romans mining for ore and became known as 'The Scowles'. In the 18th century, the incumbent landowner enhanced the feature with footpaths and rock-cut steps into a 'quaint' attraction to impress visitors with, following the fashion of the time for landscaped gardens with suggestions of antiquity and rural simplicity—grottoes, temples, footbridges, and resident hermits. By the start of the 20th century, Puzzlewood became open to the public, for a modest donation at the gate. It wasn't until late in that century that it became a fully fledged tourist attraction—its success largely due to its use as a popular TV and film location. Among the many productions that have been filmed there, there is a significant emphasis on Fantastika: *Star Wars: The Force Awakens* (2015); *Jack the Giant Slayer* (2013); *Snow White and the Huntsman* (2012); *The Huntsman: Winter's War* (2016); *The Secret Garden* (2020); *Doctor Who* (2005–2022); *Merlin* (2008–2012); *Atlantis* (2013–2015); *Da Vinci's Demons* (2013–2015), and others. Kathleen Kennedy, *Star Wars* Producer and President of Lucasfilm, described it as 'the most magical forest on the face of the earth!' In the café, posters from the most significant productions and a replica sword—a prop from one of the films—are proudly on display. The website, trail guides, and signage also encourage making associations with the associated Fantasy texts, and more 'traditional' folklore (although part of the fascination is seeing how these two frames are not mutually exclusive, but bleed into one another in interesting ways). A popular pastime of visitors—apart from the ubiquitous selfies and snaps—is the hunting down of specific locations featured in particular scenes in the more famous movie productions—especially *Star Wars*. Visitors use their phones to compare stills with the actual location in front of them, and often exclaim in delight when they have identified one. This is then photographed with friends or family members in situ, sometimes mirroring the poses of their favourite characters from the franchises. Self-confessed film fanatic Thomas Duke, instigator of the 'Stepping into Film' Instagram project, reflected upon his visit to Puzzlewood on a blog featured on their website:

> From lightsaber-wielding Jedis to giants, sorcerers and a flying blue time machine, these woods have seen it all and I was so excited to come and explore Puzzlewood. And it didn't disappoint. It was a rather surreal experience in terms of environment and film content—so much has been filmed at Puzzlewood and it was a perfect place in which to immerse myself deep within its otherworldly foliage and deep-rooted nature.
>
> (Puzzlewood, 2022)

The way visitors like Thomas Duke use a particular 'landscape' frame through which to view a real-world location is a contemporary updating of the creation of 'landscape painting' from the 16th century onwards (Owens, 2020). The compositional similarities between a 'landscape' (painting) and the classic letterbox movie frame (often cropped to fit the proportions of a large screen TV better) is no coincidence, but the culmination of centuries of societal cultivation. Via the conventions of landscape painting and photography (and now our own camera-phones), our eyes are trained to see the natural world in this rectangular way (Berger, 1972). Early drawing frames helped the artist to 'frame' the landscape, and to seek out the most harmonious composition, with proportions aligning with the Classical predilection for the Golden Section. Sometimes modern versions of these viewing frames can be found in the middle of the countryside—encouraging the walker to see the vista through this particular Neoclassical lens. The modern obsession with selfies is merely a new iteration of this. Something does not 'exist' unless it is framed in some way—usually digitally, and thus reduced to pixels, or to a long string of binary code, which can be stored and transmitted almost instantaneously. The real is made 'realer' by its televisual validation. Debord foresaw this in 'The Society of the Spectacle', as he comments here:

> In 1967, I showed in a book, *La Société du spectacle,* what the modern spectacle was already essentially: the autocratic rule of the commodity

> economy having attained the status of irresponsible sovereignty, and the set of new techniques of government that accompany this rule.
>
> (1990: 2)

The voyeur-consumer is sovereign (or at least made to feel they are by the multinational mega corporations who exploit them). Social media encourages our narcissism and endemic oversharing—so that we become unwittingly complicit in supplying the algorithms with valuable data, which is sold on to businesses wishing to profile us. As the 2020 Netflix documentary *The Social Dilemma* emphasised, 'If you are not paying for the product, you are the product.' And yet, despite the somewhat insidious implications of lives continually requiring digital validation (a phenomenon critiqued in Dave Egger's 2013 Facebook satire, *The Circle*), the reflected glamour of TV and film productions does have a tangible effect on the way a landscape is perceived and conserved—as Puzzlewood demonstrates. Its continuing popularity helps protect it, *and* generate revenue for both the attraction and surrounding local businesses (accommodation, restaurants, and other cross-promoted attractions). Within the Forest of Dean, Puzzlewood is part of a tourist 'ecosystem', which is generally mutually supportive. The discovery (and subsequent loss) of a hoard of Roman coins there seems a symbolic foreshadowing of the lucrative tourist revenue generated. Young visitors are encouraged to go on a hoard hunt throughout the site—spotting the plastic 'coin' signs. In doing so, they (or rather their parents) create new wealth via their visit. It is easy to become disorientated in the labyrinth of pathways. The nooks and crannies, twists and turns, certainly stir the imagination. And it is reminiscent of the Hollister, the 'Elf Garden' in **Hafnafjörður**, Iceland—a municipal garden created from a lava flow, and comprising a charming series of dells linked by winding footpaths. One could easily imagine Puzzlewood being the domain of elves (or the English equivalent). Yet there seems to be no local folklore.

However, just down the road there is a site which does have not only folklore but significant association with perhaps the most important Fantasy text of the modern age.

Lydney Park is only seven miles southwest of Puzzlewood. Within the grounds of the private estate, there is a remarkable Romano-British temple associated with the obscure deity Nodens (a Romano-British god of dreams who only crops up in two other places in the British Isles—on the Lancashire coast and on Hadrian's Wall). On a flattened hilltop deep within the estate, sequestered amid rich deciduous forest, the foundations of the dream-temple can be found. One can still walk up the steps to the inner sanctum, see the *abatons* or sleeping cells, where pilgrims would spend a night, often accompanied by temple dogs, in the hope of receiving a healing dream to be interpreted by a resident haruspex in the morning. As at the Romano-British temple of *Aquae Sulis* (modern-day Bath) just over the Severn, offerings of lead images of limbs were cast in the spring along with curses on lead scrolls. One of these *defixio*, or 'curse tablets', relates the theft of a ring:

> For the god Nodens. Silvianus has lost a ring and has donated one half [its worth] to Nodens. Among those named Senicianus permit no good health until it is returned to the temple of Nodens.

When Sir Mortimer Wheeler excavated Lydney in the 1920s this inscription caught his eye. In particular, the identified suspect, Senicianus. He had seen the name elsewhere. Eventually he made the connection. In 1785 in a field near Silchester a farmworker discovered a magnificent Roman ring. This became acquired (no one knows how) by the Vyne family in Hampshire, and they have passed on this family heirloom ever since, which has become known as The Vyne Ring. Upon it, there is an inscription in Latin: SENICIANE VIVAS IIN DE (it should have ended 'Viva in Deo'—live in God). In 1888, the owner of the property wrote about the 'Ring of Silvianus' in his history of the house, and this is how Wheeler came to know of its existence. While excavations were under way, Wheeler invited a promising young philologist at Oxford University to research the etymology of the ring. His name was J.R.R. Tolkien. Tolkien visited Lydney, and was fascinated not only by the ring—how did an owner-cursed ring end up in some farmland near Silchester, only to be discovered by a farmworker centuries after it had been lost? And more—the park had its own 'Scowles', and these, unlike Puzzlewood, had their associated folklore: it was believed the mines were haunted by a mysterious figure known as the 'Lord of the Mines' and by a race of dwarves (the Roman workings were known locally as 'Dwarf's Hill'). Tolkien dutifully wrote up his findings in an addendum to Wheeler's 1932 survey report of the site (Tolkien, 1932) And it is extremely likely that the evocative story of a cursed ring—one that was lost and found; the 'Lord of the Mines', and a race of dwarves who dwelt within, all influenced the creation of his magnum opus, *The Lord of the Rings*. Of course, Tolkien had been working on his legendarium since the First World War, and his experiences during conflict—and the loss of two of his beloved TCBS—provided perhaps the greatest modus operandi beyond his own love of languages, but clearly Lydney can lay legitimate claim to being one of the formative sites of Tolkien's epic. The figure of Nodens himself—cognate with the cthonic British deity, Nudd (father of Gwynn ap Nudd—Lord of the Wild Hunt, associated with Glastonbury Tor), and the Irish Nuada (of the silver hand), must have intrigued him too—and it is hard not to see a prototype of Lord Elrond of Rivendell in him, or Thranduil, the 'Elf King' of Mirkwood featured in *The Hobbit*.

The connection with Tolkien is only mentioned by external sites, and Lydney Park is inaccessible to the public except for three weeks in May. Yet there is mention of Tolkien on the interpretation board in situ. Wheeler's report emphasised the significance of the site—published only one year after the updated Ancient Monuments Act of 1931, it no doubt helped to protect the park from insensitive development and has done to this day. The Vyne Ring went on display in Hampshire in the 'Ring Room' in The Vyne (Mitchell, 2013), now a National Trust property, in 2013, along with a first edition of *The Hobbit*, and in the grounds, there is a 'Middle-Earth' children's playground.

For other sites associated with Tolkien's formation of Middle-Earth we need to travel farther north. Taking the Roman road of the Fosse Way from Cirencester we pass the charming Cotswold towns of Stow-on-the-Wold, and Moreton-in-Marsh. Both have Tolkien sites. At the former, St Edward's church is a predominantly Medieval structure on the site of a Saxon place of worship. Its north door (traditionally the 'Devil's Door') is framed, and almost overwhelmed by two ancient yew trees which dominate that side of the church. These are said to be the inspiration for the 'Door of Durin'—the entrance to the Mines of Moria in LOTR—and is, according to the Stow Civic Society website, 'the most photographed door in the Cotswolds'. On a nearby yew tree there is a laminated notice with a QR code inviting you to guess: 'Are the yew trees holding up the church, or vice versa?' and to make a donation to help preserve the building. Although not explicitly mentioning the Tolkien connection, the sign is clearly placed to capitalise upon the many visitors who visit it for that reason, and it is using the synthesis of myth and history to raise funds. A short ride up the road in Moreton-in-Marsh, which straddles the Fosse Way and sports a handsome dragon-slaying knight on its central war memorial, there is The Bell Inn—said to be the inspiration for The Prancing Pony, as a blue plaque acknowledges. Inside, a snug bar has been made into something of a 'Tolkien room', with photographs of the professor in his elder years, and a map of Middle-Earth adorning the wall. Clearly, here, the associations are good for business.

But we must go a little farther up the Fosse Way to find the 'ground zero' of Tolkien's Middle-Earth.

Birmingham

The parents of John Ronald Reuel Tolkien, Arthur and Mabel Tolkien, were originally from Birmingham, but moved to Bloemfontein, South Africa, to further Arthur's banking career. And it was here, in the 'City of Roses' (or 'Place of Cheetahs', in the city's Sesotho name, Mangaung) on 3 January 1892, that the 'Author of the [20th] Century', was born.

The infant Tolkien spent his first three years in this hot, dusty 'fountain of flowers', before his mother took him and his brother Hilary to visit her parents back in Kings Heath, Birmingham. While there, Mabel heard the terrible news: her husband had died of rheumatic fever. And so the Tolkien family settled once more in England's 'second city', moving between nine homes from 1895 to 1911. Although he never returned there once he left for Oxford University, Tolkien always thought of himself as a 'Birmingham Man'. The first house the Tolkien family moved to was a hamlet close to Sarehole Mill, which at the time was on the edge of the city and abutting the Worcestershire countryside. This proved something of a eucatastrophe after the tragic loss of the patriarch, for the two young siblings found refuge, solace, and adventure in Sarehole Mill. It was to prove foundational to Tolkien's legendarium—a cornerstone of The Shire; with the grumpy miller, whom they nicknamed naughtily 'The White Ogre' at the time becoming Ted Sandyman in LOTR. The significance of this burgeoning mythscape on the young Tolkien was captured in his early poem, 'The Cottage of Lost Play', which is perhaps a summation of all of his childhood play spaces rather than one in particular, but it shows how his mythopoeia took root in the soil of the Midlands. From a childhood perspective, the millpond—now choked with duckweed—became a great lake, and one willow tree in particular the infant Tolkien grew very fond of, to the point that when he discovered it cut down one day, he was devastated—a loss which seemed to feed directly into Treebeard's solastalgia at the wilful destruction of trees in Fangorn forest by Saruman's minions. Near Sarehole Mill is Moseley Bog—formerly the site of a second, larger millpond, and now a nature reserve managed by the Birmingham and Black Country Wildlife Trust, whom I met undertaking conservation work when I visited it in the summer of 2022. The rare bogland habitat is home to buzzards, great spotted woodpeckers, and pipistrelle bats, and is an important archaeological site—with two burnt mounds (sites where some form of 'sweat lodge' took place—although whether for ceremonial or more mundane ablutionary purposes it is hard to say). For Tolkien, it provided inspiration for 'The Old Forest' in LOTR—the home of Tom Bombadil, who is a curious outlier in the narrative, a fossil trope from a different genre: Fairy Tale, rather than Epic. Visiting Moseley Bog and Joy's Wood, it is easy to see how such a place must have nourished Tolkien's imagination—it feels slightly 'wilder' and less obviously anthropocentric than Sarehole Mill. The winding, forking pathways suggest alternative narrative threads via their arborescence. At one point, there is a clear fork and one is forced to choose which path to take—perhaps the Tolkien siblings chose one each and regrouped further on to relate their separate adventures, like the sundered hobbits in FOTR? We can only speculate, but one thing is more certain: the importance of such play spaces to developing minds and bodies, as Macfarlane and others have commented (Opie, 1951; Macfarlane, 2015; Krznaric, 2020). In 1900, the family moved to a new home in Moseley, and John Ronald and Hilary started taking the tram to King Edward's School. Later, they would move to Edgbaston—dominated by the two towers of Perrot's Folly and the Edgbaston Waterworks: direct visual inspiration for Minas Morgul and Minas Tirith in *The Two Towers*. Other sites in the city (The Oratory; Plough and Harrow; University of Birmingham; Central Library) are highlighted in the 'Tolkien Trail' leaflet, developed with The Birmingham Tolkien Group.

Exploring Tolkien's old 'neck of the woods' in person made him more relatable, if anything. Imagining him and his brother Hilary playing around there as boys reminded me of my own childhood adventures and how vital such places are—as dreaming spaces. By its association to Middle-Earth, these parts of Birmingham become 'enchanted' (and are protected). Unlike many green spaces in urban areas, Sarehole Mill and Moseley Bog were not festooned in litter. It was impressive to see how well maintained the nature reserve in particular was—clearly cherished by local people, under the good stewardship of the Wildlife Trust it has become an important (and rare in the area) fenland habitat. One can imagine how Tolkien, and perhaps Tom Bombadil himself, would have approved.

And yet, despite the small exhibition at Sarehole Mill, and the trail guide, Birmingham's links to Tolkien are not as celebrated as they should be—the city is not promoted as 'Tolkien's City', or 'The Capital of Middle Earth', in the way other cities with literary links are (Jane Austen's Bath; Dickens' London, etc.). A Middle Earth Festival used to be held at Sarehole Mill, run by volunteers, but this has ceased. Perhaps the new mega-budget Amazon loose reworking of *The Silmarillion, The Lord of the Rings: The Rings of Power* (2022–) will see a surge of new interest in the author, which tourist boards may seek to exploit. For now, it is evident to see how Tolkien's associations with Sarehole Mill and Moseley Bog have helped to preserve them—indeed Tolkien himself made a generous donation to help the restoration of the mill in 1960. It was clear how much the place meant to him.

Alderley Edge

Alderley Edge is a wide red sandstone escarpment and horseshoe-shaped hough dominating its corner of the Cheshire landscape, with sweeping views (on a clear day) from its crags 215 metres above sea level across the county's plains, and towards the Peak District. Managed by the National Trust, it is a popular place to visit from Wilmslow, Macclesfield, Manchester, and further afield, receiving 300,000 visitors a year (NT). With its Bronze and Roman mines (and recently discovered cobalt mine, National Trust, 2022), and rich woodland, it is a Site of Special Scientific Interest.

With its evocative features—Stormy Point, Goldenstone, Druid's Stone, Castle Rock, and the Wizard's Well, it is also a site of special author interest—in particular to Alan Garner, who has lived within a mile of the Edge for the majority of his ninety-plus years. His intimate knowledge of the area, akin to Nan Shepherd's 'deep mapping' (2011 [1977]) has informed his oeuvre, especially The Weirdstone Trilogy: *The Weirdstone of Brisingamen* (1960); *The Moon of Gomrath* (1963); and *Boneland* (2012), which draw heavily upon the psychogeography of Alderley Edge and its environs. Local landmarks and places featured in the books include:

- Alderley Edge
- St. Mary's Clyffe
- The Edge
- Castle Rock
- Holy Well
- Stormy Point
- Iron Gates
- Druid Stones
- Old Quarry
- Golden Stone
- The Wizard Inn
- West Mine
- Highmost Redmanhey
- Radnor Wood
- The Parkhouse
- Dumville's Plantation
- Monks Heath
- Sodger's Hump
- Bag Brook
- Marlheath
- Capesthorne Hall
- Redesmere
- Thornycroft Hall
- Pyethorne Wood
- Gawsworth
- Danes Moss
- Macclesfield Forest
- Shuttlingsloe
- Piggford Moor
- Clulow Cross

One can see why such evocative placenames appealed to Garner, and why this helped to shape such a chorographical work. One could argue that many Fantasy narratives have this quality, but Garner was keen to situate his Fantasy in a real landscape, as he discussed in a 1968 interview: 'If we are in Eldorado, and we find a mandrake, then OK, so it's a mandrake: in Eldorado anything goes. But, by force of imagination, compel the reader to believe that there is a mandrake in a garden in Mayfield Road, Ulverston, Lancs, then when you pull up that mandrake it is really going to scream; and possibly the reader will too' (cited in Philip, 1981: 25). For Garner, the 'affect' of Fantasy is more tangible in a mundane context but is constrained by the actuality too—creating an existential tension, which struggles to have real traction in most Secondary World narratives. It is not 'anything goes'. As with Garner's increasingly sparse and cryptic prose style (which took a ludic, heteroglossic turn in his Booker Prize-shortlisted *Treacle Walker*), the Garnerverse flourishes through constraint. It is as though the author has chosen to embody Hamlet's claim: 'I could be bounded in a nutshell, and count myself a king of infinite space' (Act 2, Scene 2), deciding to embrace the ensuing 'bad dreams' which the Prince of Denmark is wary of. Garner's protagonists are certainly plagued by them—hallucinatory, neurotic, even psychotic episodes manifest even in the first two of the trilogy (which were marketed on publication for younger readers initially), often in terrifying episodes, such as the vision of the Morrigan from Stormy Point in *The Moon of Gomrath*. The mental health, or lack of, of the main protagonist, Colin—now a professor of astrophysics working at Jodrell Bank—in the late addition to the trilogy, *Boneland,* is key to the plot: is Susan a symptom of his illness, or did she really exist (before mysteriously disappearing)? And can she somehow still exist in the firmament which Colin scrutinises, a personification of 'lookback time' (Oxford Reference, 2022), as astronomers call the light of long-dead stars? It is as if the landscape in Garner's work also serves as a mindscape. The two are inseparable, almost claustrophobically so, as the documentary, *Edge of the Ceiling* (1980), implicates—evoking both the atmosphere of the novels and Garner's somewhat bosky deep process. Stricken with an illness, the young Garner was forced to convalesce in his Cheshire bedroom where strange visions appeared on the ceiling, in a similar way to Gloucestershire writer Laurie Lee's own childhood illness epiphany (Lee, 1959). The almost shamanic flights of fancy he experienced seemed to feed into his later works, which often dramatise fugue states—especially *Red Shift* (1973) and *Treacle Walker* (2021). Yet, mostly bound to earth, his young protagonists, Colin and Susan, in the first two Weirdstone books are physically forced to transition into and out of the underworld labyrinth beneath Alderley Edge, which bleeds from the actual mines into an Annwn-like space of supernatural beings and legendary figures (Garner draws upon the local—but also widely-found—folktale of the man going to market to sell his horse being met by a mysterious figure by Wizard's Well, who offers to buy the horse. At first the man refuses, but upon returning from market empty-handed, accepts, and is taken into a secret passage beneath the Edge where a King and his Knights sleep—here he is offered all the gold he can carry. He leaves more than satisfied, but try as he may, he never can find the entrance to this subterranean treasure house again). There is a pronounced verticality and vertiginousness in Garner's work—no doubt influenced by the steep slopes and deep mines of The Edge, and reaching its apogee in Treacle Walker, where the boy protagonist travels up the chimney in his cottage, using it like the World Tree or Shaman's Ladder to traverse interdimensional space.

The association with Garner's work is briefly mentioned in the official signage greeting visitors to the site, next to the (closed in 2022) 'Wizard's Tea Room', but the trails evoke the local folklore in more tangible ways ('Wizard's Wander'; 'Miners' Trail', etc). The trails are well maintained by the National Trust, and the site is clearly loved and looked after.

Visiting Alderley Edge after reading *Moon of Gomrath* and *Boneland* earlier in 2022 was certainly resonant. One could see how the site stirred Garner's imagination. Also, while there, I stumbled upon the location of a short film on YouTube (Saunders, 2022) about a 'genuine' pixie-hunter—Erwin Saunders—and his 'First Sighting of a Pixie'. Saunders ensconces himself in a quiet corner of The Edge, a place he identifies as a likely site for spotting the little folk. Drawing upon the found footage technique as used in *The Blair Witch Project* (1999) to create verisimilitude, Saunders acts the bumbling amateur eccentric, but slyly slips in some CGI of a pixie just when you least expect it. Stumbling upon this location (which I had not realised was filmed at The Edge) made me perceive it in a different way, even though I knew it was a hoax. For a little while I sat in the same spot as Saunders and secretly hoped a pixie would appear to me—even a CGI one. Later, tracking down the Wizard's Well took persistence—on a hot day it was even a kind of quest itself. After a couple of attempts which left me retracing my steps to Stormy Point, I asked directions and was pointed to another two wells—both of which flowed better than the Wizard's Well, which I eventually found but was disappointed to find was more or less stagnant. Having visited it in the late 80s, when I remember it *was* flowing, I felt a little like the protagonist in the folk tale who is unable to find the secret entrance upon his return. The famous inscription above the well was barely legible—and here I also relied upon memory to fill in the gaps: 'Drink of this and take thy fill for the water falls by the Wizhard's [sic] will.' This was reputedly created by local stone mason Robert Garner, Garner's great-great-grandfather, in the 19th century. And in his oeuvre Garner has continued this family-line of mythopoeia, creating an enchantment about Alderley Edge, which has drawn countless visitors, and ensured its conservation ever since.

Lud's Church

Near to Alderley Edge, over the border into Staffordshire at Gradbach, there is a spectacular millstone grit gorge, which has been associated with one of the taproot texts of English Fantasy, *Sir Gawain and the Green Knight*. The result of an underground river system collapsing, Lud's Church received its name from Lollard dissenter, Walter de Ludank or Walter de Lud-Auk, who was captured here at one of their meetings. Followers of early church reformer John Wycliff, the Lollards were supposed to have used the gorge as a secret place of worship during the early 15th century, as a refugium from persecution for their religious beliefs. The narrow gorge would have been particularly conducive to oratory. At one point a wooden effigy, a figurehead from the ship *Swythamley*, formerly stood in a high niche above the chasm. It was placed there by Philip Brocklehurst, then the landowner, around 1862. Called 'Lady Lud', it was supposed to commemorate the death of the daughter of a Lollard preacher, although it is tempting to imagine it being a personification of the genius loci. Approached from the far end of the Dane Valley, Lud's Church took some finding—there was only a small wooden sign pointing the way *after* one had located the correct junction in the forest. For somewhere relatively obscure, it was surprisingly popular—indeed, there was something of a bottleneck of people awaiting to enter or exit the gorge when I visited (no more than dozen, but with children and dogs in a gorge it quickly became quite noisy at times). Nevertheless, when one had scrambled down into the depths of the chasm, flooded but fortunately lined with duckboards, the affect of the place became tangible. Ahead, light could be glimpsed between the ever-narrowing fissures—tantalisingly just out of reach. More than the text of the poem, it was David Lowery's bold cinematic re-imagining of it, *The Green Knight* (2020), which influenced my perception of this Fantasy site. I found myself wanting to hear the chilling sound of axe-grinding. I also wanted to experience the eco-epiphany that seems to be alluded to in the final climactic confrontation in the Green Chapel, when Gawain, after all of his travails, finally makes his destination and doom. He sits cross-legged and Siddhartha-like at the foot of the apparently dormant Green Knight—who had almost become incorporated once more into the sylvan realm, a Bodhi Tree of the North—and awaits his fate. At dawn the Green Knight stirs awake, and the Beheading Game begins, or rather concludes, in earnest. Gawain at first flinches. But then, ashamed, he pulls off the green garter given to him by Lady Bertilak. The Green Knight kneels and responds: 'Well done, my brave knight. Now, off with your head.' Whether Gawain survives the killing stroke is left open in an ending that has been much debated (Robinson, 2021). Other sites have been suggested for the location of the Green Chapel, but as we are talking about a Medieval Fantasy, it is perhaps a moot point—akin to trying to locate King Arthur's 'actual' grave (several exist). Lud's Church fits the bill perfectly—it 'performs' to our expectations, as a place of mystery, sinister potential, and revelation. There is a tangible sense of the Sublime about it, in direct compliance with Burke's ideation: 'Whatever is in any sort terrible or is conversant about terrible objects or operates in a manner analogous to terror, is a source of the sublime' (1757). What is more terrifying than nature given agency—and an axe?

The Road Goes Ever On

There are other sites I visited on my field research in the summer of 2022—Snowdonia and Anglesey, Wales; The Old Library, Trinity College, Dublin; Newgrange and the Boyne Valley; Queen's University, Belfast; Game of Thrones Studio, The Dark Hedges, Giant's Causeway, and The Causeway Coastal Route, Northern Ireland; Rathlin Island, County Antrim; Drumnaph Nature Reserve; Teltown and Tara, County Meath; and various sites in Iceland—but for now I hope this brief survey gives a sense of the affect of such places, and how Fantasy narratives influence our perception, appreciation, and treatment of them.

Conclusion

The Le Guin Index and Other Interventions

Having conducted our survey of the environmental aspects of Fantasy, it is time to draw some conclusions and to offer some interventions—in keeping with the idea that Fantasy is not a passive thing, a mere distraction in the general sideshow of 'culture', but one that can be activistic, raising awareness and shifting perceptions about the natural world.

Just because a Fantasy features a natural setting doesn't make it 'green'—it needs to be more than a pleasant backdrop. Ideally, the very 'nature' of nature needs to be questioned, or at least its continuation. Its survival and health should not be a given. This is increasingly relevant and resonant in the shifting baseline of the Climate Crisis. As the nature of the biosphere changes, so too must the nature of Fantasy, for nothing is in isolation—even the apparent refugia of Fantastika. The production and consumption of Fantasy and its cousins take place in *this* world, wherever it is set.

Since the existence of taproot texts like Gilgamesh came to light, early modes of Fantasy have explored the tension between the human and their storied environment—one which they must learn to share with the more-than-human and supernatural.

In many of these narratives, the breaking of taboos around specific sites results in consequence—a kind of accelerated 'ecological karma' (anthropogenic Climate Change in microcosm). Through narrative compression the idea is emphasised that our actions have consequences; that the more-than-human (habitats and critters) always notices, and often suffers as a result; and ultimately, insensitive treatment of our shared biosphere results in our own suffering.

Increasingly, ecological anxiety and themes are becoming foregrounded and explicit in Fantasy (and its close cousins, speculative fiction and ecofiction).

Fantasy, in its broadest sense—as a technique for imagining other modalities—is a key tool in strategies for awareness-raising, imagining alternatives, and ameliorating the challenges of the Climate Emergency, as

DOI: 10.4324/9781003463665-9

the works of Rob Hopkin, Jonathan Porritt, Roman Krznaric, Robin Wall Kimmerer, and others advocate.

Fantasy also provides an important refugium for coping with the multiple stresses of modern life—its benefit to our mental health cannot be underestimated. It celebrates alterity and plurality, and encourages a playful plasticity in our experience of reality.

Whatever the future has in store—one thing is clear: we'll still need stories, because as Emily St John Mandel's near-future post-pandemic novel *Station Eleven* (2014) points out. 'Survival is insufficient'.

For now, I will suggest the Le Guin Index, for assessing and categorising the ecological credentials of Fantasy texts, as a way of promoting and encouraging existing and future works. Other ethical aesthetics will be also considered.

Before we consider this criteria though, let us consider the environmental aspects of Le Guin's oeuvre.

Environmental Aspects of the Works of Ursula K. Le Guin

Much of Le Guin's extensive oeuvre demonstrates environmental awareness—it is there in the archipelagic setting of Earthsea (the very name of that Secondary World suggests an ecological balance) and its complex ecosystems of cultures and species, and the anthropological and ecological principles they operate by. The magical system of that world—as first explored in her early stories, 'The Word of Unbinding', and 'The Rule of Names'—enshrines ecolinguistic reciprocity as the sine qua non of its reality, as epitomised by the wizard's maxim: 'the truename is the true thing.' Discover the truename of something or someone, and you can control it. The latter is set on a small island in Earthsea, Sattins Island, 'where a careless adjective might change the weather for a week.' In 'The Word of Unbinding', a mage called Festin attempts to escape a spellbound prison by changing his shape into many forms: a whisp of smoke, a flame, a hawk, a trout. In the latter form, he hides in a cool, shadowy pool, soothed, as Le Guin describes, by 'running water, stronger than any spell of healing'. Festin eventually escapes through death and defeats his foe by summoning him there by his truename. A Lethe-like place, Festin begins to forget 'the voice of streams and the sound of rain on the leaves of the forest of life'. In 'The Finder' (*Tales of Earthsea*) we find out about the 'Immanent Grove', where the women of one island go to connect with the source of magic. Ember is the authority on it, but is initially taciturn, refusing to elucidate its Mysteries: 'You can learn about the Grove only in it and from it' (Le Guin, 2018).

Ember emphasises the unique qualities of each tree, qualities tied to its specific name: 'In the Old Speech, Ember said, each of those trees had its own name' (2018 [2001]: 600).

Medra asks: 'How far does the forest go?' Ember, echoing Buddhist teaching, gnomically replies, 'As far as the mind goes.'

Then Medra hears the leaves of the trees 'speak', and she finds that the shadows could be read. 'I am learning to read them,' she says (p. 601). This echoes the famous dictum from the Bible—'the leaves of the trees are for the healing of nations'.

In *The Word for World Is Forest* (1972), Le Guin foregrounds the eco-linguistic in the title, and the ecological throughout a plot which James Cameron remediated through his first Avatar (2009) film.

'As a fiction writer, I don't speak message, I speak story . . .' This is something Le Guin believed, even towards the end of her long and distinguished career. The statement was partly in exasperated response to school librarians and teachers asking her for books with the 'message' front-loaded. Philip Pullman, a writer very different in tone than Le Guin but no less ambitious in his grappling with big ideas about the human condition and the nature of reality, has experienced a similar problem. In discussing avoiding didacticism in his writing, Pullman reflected: 'Ideas are best conveyed by making them look not like ideas at all, but events.' Le Guin, in this slim novel first published in 1972, at the height of the Vietnam War, has tried to do just that—not by offering a polemic, but by dramatising the problem—the wearily familiar dynamic of an aggressively Capitalist colonial power subjugating and exploiting the indigenous inhabitants for the purposes of extracting the maximum amount of wealth (resources, in the form of either slave labour or mineral wealth or both), no matter the cost to the ecosystem, its biodiversity, and the lives of the Aboriginal dwellers. This rapacious pattern (the inevitable manifestation of 'progress without limits', and 'free trade', the mantra of late capitalism, Neoliberalism) has played out throughout humanity's history—although that should not make it normative. We come to think of it as the only game in town, and the way things are, when, with enough willpower, other ways are possible and other worlds. Le Guin famously challenged this hegemony in her acceptance speech for the National Book Foundation Medal for Distinguished Contribution to American Letters (2019):

> We live in capitalism, its power seems inescapable—but then, so did the divine right of kings. Any human power can be resisted and changed by human beings. Resistance and change often begin in art. Very often in our art, the art of words.

In Le Guin's story—set within the archipelago of novels exploring the Hainish universe (an advanced intellectual race who may or may not be the progenitors of humankind)—the author transposes this pattern onto a distant planet. Although it is forgivable to see the echoes of the Vietnam conflict, by setting the action on a richly tropical planet the human colonists call 'New Tahiti', Le Guin creates the cognitive estrangement which defamiliarises us and makes us see the situation anew. Here, both sides are flawed, moral dualisms become enmeshed in complexity, and the obvious empathic leap, identifying with the diminutive hirsute natives (the Athsheans as the literal 'underdogs') starts to feel uncomfortable when we see them enacting 'war crimes' against the Colonist just as heinous as those they have endured. The cycle of violence seems inexorable until the arrival of a passing ship from Earth, the Shackleton, conveying Hainish and Cetian ambassadors and an 'ansible' (interplanetary communicator) changes the rules of the game. Le Guin's father was an anthropologist, and her exposure to his discipline and careful methodology informs not only this, but much of her work—yet here it is foregrounded in the figure of Raj Lyubov, a 'spesh' (specialist) who has deep sympathies for the Athsheans after learning their language and studying their sophisticated dream-praxis. Yet although Lyubov's frustrated attempts to ameliorate the brutish treatment of the 'Creechies' (as the inhabitants are dehumanised), and wanton destruction of the sylvan biome (the extraction of the incredibly valuable timber drives the 'annexing' of the planet) are for a while the main line of desire, ultimately Lyubov is seen as an emasculated protagonist, amid the ultra-machismo of the military endeavour (which Le Guin darkly satirises: the staccato jargon, testosterone-pumped behaviour, and self-destructing madness mirroring Heller's *Catch-22*). As the viewpoint shifts to Selver, the Athshean rebel figurehead 'god', it is easy to assume we are going to fall into a classic rebellion narrative—Robin Hood in space; but Le Guin has more sophisticated fish to fry. To say more than that would involve spoilers; but the whole novel (practically a novella) is familiar to movie-goers across the world, for James Cameron's *Avatar* feels like a very loose, unofficial 'revisioning' of it. However, those expecting things to play out in a similar way will be deeply surprised. Le Guin is not one to go for the crowdpleasing payoffs. Her universe is more complex than that. In such a short 'novel', it is perhaps inevitable that the characters will seem (relatively) thinly sketched: Captain Davidson, the violent, indignant antagonist, is the least convincing; although there is nothing simple about Mr Lepennon, the Hainish visitor, or the troubled, reluctant messiah, Selver. What comes across most authentically amid the ambitious thought experiment of it all (the logical endgame of infinite growth on a galactic stage) is Le Guin's cri-de-coeur for the protection of precious habitats, the biodiversity of life they contain, and the autonomous rights of those who dwell among them. Environmental awareness was developing when she wrote the book, but it feels increasingly prescient in an age of Climate Chaos and the latest IPCC report urging governments to act now before it is too late. As such, Le Guin's book is a message in a bottle from the future (written in our past)—but one that interweaves that 'message' very skilfully into the texture of the narrative, challenging us as readers: confronting us with our assumptions and complicities. But that is to make it sound abstract and intellectual, when in essence it is an Adventure Story told at a cracking pace. To let Le Guin have the last word: 'The complex meanings of a serious story or novel can be understood only by participation in the language of the story itself. To translate them into a message or reduce them to a sermon distorts, betrays, and destroys them' (1979).

Drawing upon the examples of excellent practice in Le Guin's work at exploring and foregrounding environmental thinking, I devised in 2022 what I call 'the Le Guin Index', as a way of gauging the eco-credentials of SF and Fantasy novels. This is not meant to be prescriptive or definitive, but to encourage writers, readers, and critics of SFF to consider these crucial aspects. As the United Nations declares, 'Climate Change is the defining crisis of our age' (2015). We all have to rise to this challenge. Yet even if we choose not to, everything we create and consume is inflected by it, by what Bould calls the 'Anthropocene Unconscious' (2021). Le Guin's work provides not only 'escape, consolation, and recovery', as Tolkien limns it in his essay on Fairy Stories, but also a very contemporary and resonant route-map. Professor Rupert Read uses *The Dispossessed* (1974) as a defining example of what he describes as a 'thrutopia':

> The nearest we have to a detailed thrutopia of the imagination is Ursula le Guin's The Dispossessed. That sci-fi work brilliantly juxtaposes a superficially-attractive but dystopian world of extreme inequality and only marginal democracy (a world disturbingly like our own) with what appears at first to be an anarchist utopia. But the true genius of the novel is one's gradual realisation that utopia could never be static. Any true utopia demands being continually remade. In this way, we come to understand that any utopia has to be what I call a thrutopia, on a long-term basis. On, to be precise, a pretty much permanent basis. Thrutopia isn't second best, it's the genuine article.
>
> (2017)

Avoiding the simplistic binary extremes of utopias or dystopias, Le Guin offers us something more complex and nuanced—a scenario of gender-fluidity and geopolitical complexity.

The Dispossessed is a profoundly intelligent SF novel (which we can include here under the broader category of Fantastika) interrogating every societal norm—political structures, morality, relationships, academe, nationalism, customs, fashion, possessions, and most impressively, language. This 'ambiguous utopia' provides no easy answers—finding flaws in the paradigms presented by both planets (essentially, the Anarchists and the Capitalists). The scientist rebel who 'jumps ship', travelling between the worlds to bring 'strange news from another star', is no superhero liberator, but an intellectual uneasy in his role as reluctant emancipator. It is a masterpiece of world-building which many would-be writers (and some published writers) would do well to study. Science Fiction can risk being concept-led and didactic, but Le Guin gives the machine of her universe soul—there is a 'human' story at the heart of all this, a moving portrait of love, and a salutary cautionary tale of the Blakean 'mind-forg'd manacles' we can all end up living by. If you don't dream hard enough, you end up living in someone else's dream.

It is as a thinker, as much as a writer, that Le Guin casts a long shadow. With the courage of her convictions, she was a fierce critic of injustice and inequality, and an eloquent spokesperson for our hard-won freedoms. One of her quotes has become a touchstone for the environmental protest movement: 'We live in capitalism. Its power seems inescapable. So did the divine right of kings. Any human power can be resisted and changed by human beings. Resistance and change often begin in art, and very often in our art, the art of words.' Through her words, Le Guin will live on.

The Le Guin Index

Thus, I have applied the approach outlined above in my formation of what I call the Le Guin Index. The influential author perhaps needs no justification—in her remarkable oeuvre of novels, short stories, essays, reviews, and interviews, she has blazed a trail for a very prescient eco-political form of Fantasy—one that is sharply aware of the fault-lines of the human condition and its malleable constructions of identity. So, let us break down the criteria in the Le Guin Index.

To help us survey and assess the eco-credentials of the 'canon' (whatever we perceive that to be) I propose an index to gauge and categorise the ecological qualities of a novel). It has 5 categories from 1 Tree (Poor) to 5 (Excellent).

1. 1 Tree (POOR)
2. 2 Tree (FAIR)
3. 3 Tree (GOOD)
4. 4 Tree (VERY GOOD)
5. 5 Tree (EXCELLENT)

Such a system of categorisation is of course debatable, but can provide a useful way of quickly assessing whether a text is just superficially green, or robustly environmental. It has been used in an ongoing eco-book club, Green Reads, set up in Dorset in 2023, against a range of fiction and creative nonfiction, and has proved a useful shorthand assessment tool.

However, if we wish to seek a more robust critical framework for works of the Ecological Imaginary, then I propose the following theory.

The Biosemiotic Theory of Fantasy

Biosemiotics is an interdisciplinary research agenda investigating the myriad forms of communication and signification found in and between living systems (Else, 2010). It is thus the study of representation, meaning, sense, and the biological significance of codes and sign processes, from genetic code sequences to intercellular signalling processes to animal display behaviour to human semiotic artefacts such as language and abstract symbolic thought.

What does this constitute exactly? Literary theory traditionally operates within its own sphere (literature) and rarely expands beyond that, or connects to other spheres of meaning and activity. The Biosemiotic Theory of Fantasy suggests a joined-up approach, acknowledging that the creation and consumption of literature takes place within a biome

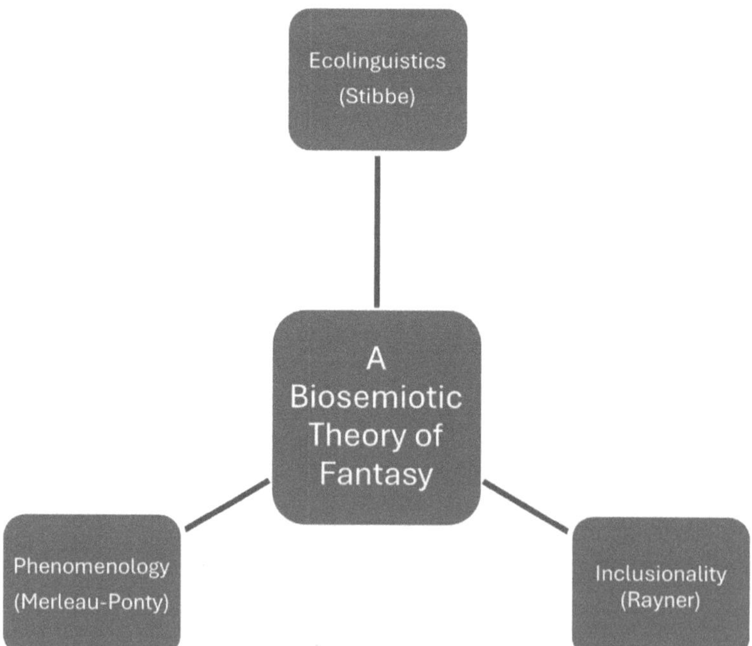

Image 8.1 A Biosemiotic Theory of Fantasy. Model. Kevan Manwaring.

(planet earth) of multiple life-forms, communities (creatives, commuters, fandom, book clubs, friendship circles, schools, colleges and universities), and biocommunities (Haraway's holobionts). There is no separation. Each influences and inflects the other. The writer is inspired by the natural world, by an article in Nature, a documentary or film, a talk, or a performance. They in turn inspire others through their writing (ideally). Then the readers, or 'prosumers' of these cultural artefacts find new ways of interacting with the natural world—after reading a chapter, watching an episode of a favourite series, playing a game, listening to a podcast, etc—they go for a walk in the local park, they visit a site associated with the text or its adaptations, they participate in an LARP event in an outdoor space, they may be inspired to participate in a local conservation programme, or attend a meeting or rally in its defence. Or maybe they just think differently about the nature on their doorstep—the birds they see from their window, the flowers growing in their garden, the quality of the air, water, and earth. It may influence their very beingness as an embodied situated being on this planet. A Fantasy text can charge all phenomena with a new, or renewed, significance. This relates directly back to the games of make-believe enjoyed as a child—Tolkien's Cottage of Lost Play. It causes a re-enchantment of the natural world, and, at best, an ontological shift. Reality is far more plastic than it may at first appear (to some), and so are we. Our identities can flicker or fully transform. The world is malleable, and we contain multitudes—the self, we discover, is a multiverse. Fantasy has enabled us to enlarge our definitions of ourselves. It has expanded both the ontological *and* Ecological Imaginary.

The Biosemiotic Theory of Fantasy proposes:

- Fantasy operates in many forms and spheres—within the cultural artefact itself, but also within interconnecting concentricities of activity and influence.
- Fantasy encourages multispecies 'communication' and/or appreciation.
- Fantasy can affect the way we perceive and treat the world.
- The affordances of Fantasy encourage neuro- and ontological plasticity; as well as a ludic and critical engagement with the consensus reality.
- Fantasy encourages a playful fluidity of self, and of beingness in the world.
- There are multiple actors and agencies involved in the production and reception of Fantasy. It is a complex ecosystem with a mycelial-like sharing of 'nutrients' and other information. The flow is seldom only in one direction. It is multi-directional, responsive, and endlessly adaptive.
- Fans of Fantasy are highly adept at noticing, interpreting, mimicking, and mutating the multifarious 'signs' of Fantasy, its codes and signals.

- The foundational principle of this theory is the idea of permeability and mutability. It is not a fixed definition, but one that will continue to adapt and evolve with further research—the greater the diversity of voices and perspectives, the better.

Drawing upon the advocacy of ecolinguistics by Naess (cited in Drengson and Inoue, 1995) and Stibbe (2021), and the Sapir-Whorf Hypothesis, the Biosemiotic Theory of Fantasy suggests that the way we communicate frames the world in which we live. The opposite of this—transactional, commercial, even militaristic—and its consequences, is seen daily in the news. Resistance is more crucial than ever, but that comes from resilience, rhizomatic networks, the sharing of resources, information, and skills, and anything that hacks the hegemonic (see 'The Rhizomatic Writer', Manwaring, 2024).

Possible critiques of this proposed theory include it is just another 'map', and that, to appropriate Auden's famous saying about poetry (1939), Fantasy does not make anything 'happen'. Perhaps not, historically, although one could argue that *all* activity connected with the production, promotion, and consumption of Fantasy is a form of 'happening' in a real-world sense. Nevertheless, to echo Auden's defence of poetry, it is a *way* of happening, and that in itself is valid. Fantasy, in the generous affordances of its myriad ontologies, offers different ways of being in the world.

Ultimately, it is the plenitudinous iterations of Fantasy themselves we return to. The Ecological Imaginary will continue to evolve, reflecting the multiple variables and shifting baselines of our calamitous present age. Its messages are increasingly urgent: that we pay attention to the natural world and honour its miraculous actuality, that we not only acknowledge the more-than-human but also respect its rights to co-exist (Macfarlane, 2025) in our shared biosphere, and that we honour the 'nature' inside all of us: that there is no absolute separation—we are hosts of biocommunities walking the planet together.

Bibliography

Abram, David (2010) *Becoming Animal: An Earthly Cosmology*. New York: Vintage.
Abram, David (2017 [1997]) *The Spell of the Sensuous: Perception and Language in a More-Than-Human World*. New York: Vintage.
Abrams, Jeffrey Jacob, Michael D. Arndt, and George Lucas (2016) *Star Wars: The Force Awakens*. Walt Disney Company.
Ackroyd, Peter (2000) *London: The Biography*. London: Chatto & Windus.
Adams, Douglas (1987) *Dirk Gently's Holistic Detective Agency*. London: William Heinemann.
Adams, Richard (1972) *Watership Down*. London: Rex Collings.
Addison, J. (1712) The Fairy Way of Writing. *The Spectator*, p. 419, 1 July, unpaginated.
Ahmed, Saladin (2013) *Throne of the Crescent Moon*. London: Gollancz.
Alain-Fournier (1913) *Le Grand Meaulnes*. Paris: Émile-Paul Frères.
Alderman, Naomi (2016) *The Power*. London: Viking.
Altes, Lisbeth Korthals (2008) "Narrative Ethics" in Herman, David, Manfred Jahn, and Marie-Laure Ryan (eds) *Routledge Encyclopedia of Narrative Theory*. London, New York: Routledge.
Alvarez-Palau, Eduard J., Dan Bogart, Max Satchell, and Leigh Shaw-Taylor (2020, 30 September) "Transport and urban Growth in the First Industrial Revolution" in *Cambridge Group for the History of Population*. Available from: https://www.campop.geog.cam.ac.uk/research/occupations/outputs/preliminary/marketaccesspresteam.pdf. Accessed 22 December 2024.
AMC (2023) *Mayfair Witches*. New York.
Aristotle, B. (1907) *The Poetics of Aristotle*. London: Macmillan.
Armitage, Simon (2008) *Sir Gawain and the Green Knight (A New Verse Translation)*. WW Norton & Company.
Atlantis (2013–2015) BBC TV Series.
Attebery, Brian (1992) *Strategies of Fantasy*. Bloomington, IN: Indiana University Press.
Attebery, Brian (2022) *Fantasy: How It Works*. Oxford: Oxford University Press.
Bach, Richard (2015 [1972]) *Jonathan Livingstone Seagull*. London: Harper Thorsons.
Ballard, J. G. (1962a) *The Drowned World*. London: Berkley Books.
Ballard, J. G. (1962b) *The Wind From Nowhere*. London: Berkley Books.
Ballard, J. G. (1964) *The Burning World* (republished as *The Drought*). London: Berkley Books.

Ballard, J. G. (1966) *The Crystal World*. London: Jonathan Cape.
Ballard, J. G. (1987) *The Day of Creation*. London: Victor Gollancz.
Barker, R. J. (2023) *Gods of the Wyrdwood*. London: Orbit.
Barnett, David (2024) *Withered Hill*. London: Canelo.
Barrie, Peter (1995 [1911]) *Peter Pan in Kensington Gardens*. London: Hodder & Stoughton.
Bartholomew, Henry (ed.) (2023) *The Unknown: Weird Writings, 1900–1937*. Handheld Press.
Barton, Anne (2017) *The Shakespearean Forest*. Cambridge: Cambridge University Press.
BB (2012 [1942]) *The Little Grey Men*. Oxford: Oxford University Press.
BBC (1984) *The Box of Delights*. London: BBC TV.
BBC (2022) "Edward Colston Statue: Protesters Tear Down Slave Trader Monument," BBC, 8 June 2020. Available from: https://www.bbc.co.uk/news/uk-52954305. Accessed 13 October 2022.
BBC (2024) "Susanna Clarke: Piranesi," *Bookclub*. Broadcast, 13 October 2024. Available from: https://www.bbc.co.uk/programmes/m0023ph6. Accessed 9 April 2025.
Beagle, Peter S. (2022 [1968]) *The Last Unicorn*. London: Gollancz.
Behrooz, Anahit (2024) *Mapping Middle-Earth: Environmental and Political Narratives in Tolkien's Cartographies*. London: Bloomsbury.
Bennett, Jane (2001) *The Enchantment of Modern Life: Attachments, Crossings, and Ethics*. Princeton, NJ: Princeton University Press.
Beresford, Elisabeth (1968) *The Wombles*. London: Puffin Books.
Berger, John (1972) *Ways of Seeing*. London: Penguin.
Bey, Hakim (1991) *The Temporary Autonomous Zone*. Pirated by HMV Þ Þ M.
Blake, William (1969 [1803]) "The Pickering Manuscript" in *Blake Complete Writings*. Oxford: Oxford University Press.
Bloom, H. (1997) *The Anxiety of Influence: A Theory of Poetry*. New York: Oxford University Press.
Booth, Wayne C. (1983) *The Rhetorics of Fiction*. 2nd edn. Chicago: The University of Chicago Press.
Borges, Jorge Luis (1945) *The Aleph*. Buenos Aries: Sur.
Borges, Jorge Luis (1946) *On Exactitude in Science*. Translated by Andrew Hurley. Available from: https://genius.com/Jorge-luis-borges-on-exactitude-in-science-annotated. Accessed 13 October 2022.
Bould, Mark (2021) *The Anthropocene Unconscious: Climate Catastrophe Culture*. London: Verso.
Boulton, James T., and Edmund Burke (1958 [1757]) *Edmund Burke: A Philosophical Enquiry into the Origin of Our Ideas of the Sublime and Beautiful*. University of Notre Dame Press.
Boyle, Danny, and Alex Garland (2002) *28 Days Later*. Fox Searchlight Pictures.
Bradshaw, Peter (2025) "David Lynch: The Great American Surrealist Who Made Experimentalism Mainstream," *Guardian*, 16 January 2025. Available from: https://www.theguardian.com/film/2025/jan/16/david-lynch-the-great-american-surrealist-who-made-experimentalism-mainstream. Accessed 8 April 2025.
Britannica. "Gaia Hypothesis." Available from: https://www.britannica.com/science/Gaia-hypothesis. Accessed 13 October 2022.
Buck-Morss, Susan (1991) *The Dialectics of Seeing: Walter Benjamin and the Arcades Project*. Cambridge, MA: MIT Press.

Bulgakov, Mikhail (1967 [1924–1940]) *The Master and Margarita*. London: Collins Harvill.
Bulgakov, Mikhail (1968 [1925]) *The Heart of a Dog*. New York: Harcourt Brace.
Bunting, Basil (1966, January) *Briggflatts: An Autobiography*. Featured in *Poetry*, vol. 107, p. 4.
Burl, Aubrey (2002) *Prehistoric Avebury*. 2nd edn. New Haven and London: Yale University Press.
Butcher, Jim (2000) *The Dresden Files: Storm Front*. New York: Penguin Putnam.
Butler, Octavia E. (2019 [1993]) *Parable of the Sower*. New York: Headline.
Cabell, James Branch (1919) *Jurgen: A Comedy of Justice*. London: The Bodley Head.
Caesar, Julius (trans. 1982 [50 BCE]). *The Conquest of Gaul*. Translated by Jane Gardner and S. Hanford. London: Penguin.
Caldecott, Moyra (1977a) *Temple of the Sun*. London: Rex Collings.
Caldecott, Moyra (1977b) *The Tall Stones*. London: Rex Collings.
Caldecott, Moyra (1978) *Shadow on the Stones*. London: Rex Collings.
Callender, Kacen (2020) *Queen of the Conquered*. New York: Orbit.
Campbell, Joseph (2008 [1948]) *The Hero With a Thousand Faces*, vol. 17. New World Library.
Carpenter, Humphrey, and Christopher Tolkien (1995) *The Letters of JRR Tolkien*. London: HarperCollins.
Carpenter, Stephen, Jim Kouf, and David Greenwalt (2011–2017) *Grimm*. NBC.
Carr, J. L. (1980) *A Month in the Country*. Hassocks: Harvester Press.
Carroll, Lewis (1894) *Sylvia and Bruno Concluded*. New York, London: Macmillan.
Carson, Rachel (1962) *Silent Spring*. New York: Houghton Mifflin.
Cascone, S. (2016) "How New York's Subway Map Was Reimagined for Women," *Artnet*, 12 October 2016. Available from: https://news.artnet.com/art-world/rebecca-solknit-women-new-york-subway-map-697910. Accessed 3 August 2021.
Casey, Edward (1988) *The Fate of Place*. Oakland: University of California Press.
Chabon, Michael (2008) *Wonder Boys*. London: Fourth Estate.
Chakraborty, S. A. (2017) *The City of Brass*. London: Harper Voyager.
Chakraborty, S. A. (2019) *The Kingdom of Copper*. London: Harper Voyager.
Chakraborty, S. A. (2021) *The Empire of Gold*. London: Harper Voyager.
Chester, Nicola (2021) *On Gallows Down: Protest, Place and Belonging*. London: Chelsea Green.
Chesterton, G. K. (1908) *The Man Who Was Thursday*. London: J.W. Arrowsmith.
Cho, Zen (2016) *Sorcerer to the Crown*. London: Pan.
Cho, Zen (2019) *The True Queen*. London: Pan.
Clapp, Susannah (2021) "How Tove Jansson's Love of Nature Shaped the World of the Moomins." Available from: https://www.theguardian.com/books/2021/jun/05/tove-jansson-moomins-woman-fell-in-love-with-an-island-summer-book. Accessed 4 April 2024.
Clare, John (1820) "The Badger." Originally published in *Poems Descriptive of Rural Life and Scenery*. Available from: https://www.poemhunter.com/poem/badger/. Accessed 1 April 2025.
Clark, P. Djeli (2021) *A Master of Djinn*. London: Orbit.
Clarke, Lindsay (2004) *Imagining Otherwise*. Green Spirit Pamphlet No. 6. Totnes: Green Books.
Clarke, Susanna (2020) *Piranesi*. London: Bloomsbury.

Clement, Jermaine, and Taika Waititi (2014) *What We Do in the Shadows*. NZ: Resnick Interactive Development.
Clute, John, and John Grant (eds) (1999) *The Encyclopedia of Fantasy*. London: Macmillan.
Clute, John, and Peter Nichols (1993) *The Encyclopedia of Science Fiction*. London: Orbit.
Coates, Ta-Nehisi (2020) *The Water Dancer*. London: Penguin.
Cohen, Jeffrey Jerome (1996) *Monster Theory: Reading Culture*. University of Minnesota Press.
Cooper, Susan (2023 [1973]) *The Dark is Rising*. London: Puffin.
Coverley, Merlin (2018) *Psychogeography*. Harpenden: Oldcastle Books.
Crawford, Jason (2020) "The Trouble with Re-Enchantment." *LA Review*, 7 September 2020.
Creamer, Ella (2024) "A Genre of Swords and Soulmates: The Rise and Rise of 'Romantasy' Novels," *Guardian*, 2 February 2024. Available from: https://www.theguardian.com/books/2024/feb/02/romantasy-literary-genre-booktok. Accessed 8 April 2025.
Crenshaw, Kimberlé W. (2017) *On Intersectionality: Essential Writings*. New York: The New Press.
Cresswell, Tim (2014) *Place: An Introduction*. Chichester: John Wiley & Sons.
Crook, Mackenzie (2019) *Worzel Gummidge*. London: BBC TV.
Crowley, John (1981) *Little, Big*. New York: Bantam Books.
Curry, Patrick (1998) *Defending Middle Earth: Tolkien, Myth & Modernity*. London: HarperCollins.
Dahl, Roald (1964) *Charlie and the Chocolate Factory*. London: George Allen & Unwin.
Dahl, Roald (2022 [1970]) *Fantastic Mr Fox*. London: Puffin.
Daiches, David (1965) "Introduction" in *Wuthering Heights*. London: Penguin.
Dames, Michael (1976) *The Silbury Treasure*. London: Thames & Hudson.
Dames, Michael (1977) *The Avebury Cycle*. London: Thames & Hudson.
Davidson, Peter (2005) *The Idea of North*. London: Reaktion Books.
de Bodard, Aliette (2015) *The House of Shattered Wings*. New York: Roc.
De Lint, Charles (1984) *Moonheart*. New York: Ace Books.
De Lint, Charles (1986) *Yarrow: An Autumn Tale*. New York: Ace Books.
De Lint, Charles (1988) *Greenmantle*. New York: Ace Books.
De Lint, Charles (1992) *Spiritwalk*. New York: Tor.
De Lint, Charles (1995) *The Wild Wood*. New York: Tom Doherty.
De Lint, Charles (2000) *Forests of the Heart*. New York: Tom Doherty.
Dean, Pamela (1991) *Tam Lin*. New York: Tor.
Debord, Guy (1954) *Potlatch*. Paris: Letterist.
Debord, Guy (1990) *Commentaries on the Spectacle Society*. London: Verso.
Debord, Guy (2012 [1967]) *Society of the Spectacle*. Bread and Circuses Publishing.
Deleuze, Gilles, and Félix Guattari (1988) *A Thousand Plateaus: Capitalism and Schizophrenia*. London: Bloomsbury.
Dery, Mark (1994) "Black to the Future: Interviews With Samuel R. Delany, Greg Tate, and Tricia Rose" in *Flame Wars: The Discourse of Cyberculture*. Durham and London: Duke University Press.
Devaney, S. (2021) "Emma Watson & Reni Eddo-Lodge Need Your Help to Rename London's Tube Map With Brilliant Women," *Vogue*, 22 July 2021. Available from: https://www.vogue.co.uk/news/article/emma-watson-rename-female-london-tube-map. Accessed 13 October 2022.

Dickinson, Emily (1876). Letter to Higginson, 1876: L 459a. Available from: https://archive.emilydickinson.org/correspondence/higginson/l459a.html Accessed 12 June 2025.

Doctor Who (1963–1989, 2005) BBC TV Series.

Donaldson, Julia, and Axel Scheffler (2017) *The Gruffalo*. London: Macmillan.

Donaldson, Stephen (1993) *The Chronicles of Thomas Covenant the Unbeliever*. London: HarperCollins.

Drabble, Margaret (1985) *The Oxford Companion to English Literature*. London: Guild Publishing.

Drengson, Alan, and Yuichi Inoue (eds) (1995) *The Deep Ecology Movement: An Introductory Anthology*. Berkeley, CA: North Atlantic Books.

Dunsany, Lord (1905) *The Gods of Pegāna*. London: Elkin Matthews.

Dunsany, Lord (1924) *The King of Elfland's Daughter*. New York: G.P. Putnam's Sons.

Eades, G., and Y. Zheng (2014) "Counter Mapping as Assemblage: Reconfiguring Indigeneity" in Doolin, B., et al. (eds) *IS&O 2014, IFIP AICT 446*, pp. 79–94. Available from: https://link.springer.com/content/pdf/10.1007%2F978-3-662-45708-5_6.pdf. Accessed 13 October 2022.

Eco, Umberto (1983) *The Name of the Rose*. New York: Harcourt.

Eggers, Dave (2013) *The Circle*. New York: Knopf.

Eglington, T. C., and Simon Davis (2023) *Thistlebone: Book One*. Oxford: Rebellion Publishing.

Ekman, Stefan (2013) *Here Be Dragons: Exploring Fantasy Maps and Settings*. Wesleyan University Press.

Elgin, Don D. (1985) *The Comedy of the Fantastic: Ecological Perspectives on the Fantasy Novel*. Westport, CT: Greenwood Press.

Eliot, T. S. (1943) *The Four Quartets*. New York: Harcourt.

Elkin, Lauren (2016) *Flâneuse: Women Walk the City in Paris, New York, Tokyo, Venice, and London*. New York: Farrar, Straus and Giroux.

Elkins, J., & Morgan, D. (eds) (2009). *Re-Enchantment* (1st ed.). New York: Routledge. https://doi.org/10.4324/9780203891667

Ellison, Harlan (1969) *A Boy and His Dog*. London: New Worlds.

Else, Liz (2010) 'Biosemiotics: Searching for meanings in a meadow'. *New Scientist*, 18 August, 2010. Available from: https://www.newscientist.com/article/mg20727741-200-biosemiotics-searching-for-meanings-in-a-meadow/ Accessed 13 June 2025

Ende, Michael (1983 [1979]) *The Neverending Story*. London: Allen Lane.

Eshun, Ekow (2022) *In the Black Fantastic*. London: Hayward Gallery.

Evaristo, Bernardine (2019) *Girl, Woman, Other*. London: Hamish Hamilton.

Farley, Paul, and Michael Symmons Roberts (2012) *Edgelands: Journeys into England's True Wilderness*. London: Random House.

Fimi, Dimitra (2009) *Tolkien, Race and Cultural History: From Fairies to Hobbits*. Basingstoke: Palgrave Macmillan.

Finley-Day, Gerry (1979) *Disaster 1990!* London: 2000AD.

Fisher, Mark (2016) *The Weird and the Eerie*. London: Repeater Books.

Flaherty, Alice W. (2005) *The Midnight Disease: The Drive to Write, Writer's Block, and the Creative Brain*. New York, NY: Mariner Books.

Flynn, Johnny, and Robert Macfarlane (2021) *Lost in the Cedar Wood*. London: Transgressive Records.

Forster, E. M. (1978 [1927]) *Aspects of the Novel*. London: Pelican.

Forster, John (1872) *The Life of Dickens*. London: Chapman and Hall.

Foucault, Michel (1997) "Of Other Spaces: Utopias and Heterotopias' by Michel Foucault" in Leach, Neil (ed.) *Rethinking Architecture: A Reader in Cultural Theory*. NYC: Routledge.
Fowler, Corinne (2019) *Colonial Countryside*. Leeds: Peepal Tree Press.
Gaiman, Neil (1990) *The Books of Magic Omnibus 1*. Burbank, CA: Vertigo.
Gaiman, Neil (1996) *Neverwhere*. London: BBC Books.
Gaiman, Neil (1999) *Stardust*. Burbank, CA: DC Comics.
Gaiman, Neil (2004 [2001]) *American Gods*. London: Headline.
Gandy, Matthew (2018) "Urban Nature and the Ecological Imaginary" in *In the Nature of Cities*. Routledge.
Gardner John (1989 [1971]) *Grendel*. New York: Vintage.
Garner, Alan (1960) *The Weirdstone of Brisingamen*. London: William Collins.
Garner, Alan (1963) *The Moon of Gomrath*. London: William Collins.
Garner, Alan (1967) *The Owl Service*. London: William Collins & Sons.
Garner, Alan (1973) *Red Shift*. London: William Collins.
Garner, Alan (2012) *Boneland*. London: Fourth Estate.
Garner, Alan (2021) *Treacle Walker*. London: Fourth Estate.
Garth, John (2003) *Tolkien and the Great War: The Threshold of Middle-Earth*. New York, NY: Houghton Mifflin.
Gerdie, Alida, Anthony Nanson, and Edward Schieffelin (2014) *Storytelling for a Greener World: Environment, Community, and Story-Based Learning*. Stroud: Hawthorn.
Gilbert, Zoe (2022) *Mischief Acts*. London: Bloomsbury.
Gilliam, Terry (1991) *The Fisher King*. Culver City, CA: Tri-Star Pictures.
Gilliam, Terry (2009) *The Imaginarium of Doctor Parnassus*. Infinity Features.
Glover, Julian (2008) "Julian Glover Reads Beowulf," *Youtube*, 19 June 2008. Available from: https://www.youtube.com/watch?v=IR464WBmA2s. Accessed 1 April 2025.
Goldman, William (1999) *The Princess Bride*. London: Bloomsbury.
Gong, Chloe (2020) *These Violent Delights* and *These Violent Ends*. London: Hodderscape.
Goyer, David S. (2013–2015) *Da Vinci's Demons*. BBC/Phantom Four.
Graham, Gordon (2007) *The Re-enchantment of the World: Art versus Religion*. Oxford: Oxford University Press.
Granada (1980) "Edge of the Ceiling, Granada TV." Available from: https://youtu.be/6D6_kO_4WqM. Accessed 8 September 2022.
Graves, Robert (1961) *The White Goddess*. London: Faber and Faber.
Gray, Alasdair (1981) *Lanark: A Life in Four Books*. Edinburgh: Canongate Press.
Grierson, Jamie (2024) "Three-Quarters of Children Want More Time in Nature, Says National Trust," *Guardian*, 1 April 2024. Available from: https://www.theguardian.com/environment/2024/apr/01/over-three-quarters-children-want-more-time-nature-national-trust?CMP=share_btn_url. Accessed 1 April 2024.
Grimmer, R. (2016) "Rethinking Genre: Genre as a Tool for Writers throughout the Writing Process," *Writing in Practice*, 2. Available from: https://www.nawe.co.uk/DB/current-wip-edition/articles/rethinking-genre-genre-as-a-tool-for-writers-throughout-the-writing-process.html. Accessed 22 June 2017.
Guanio-Uluru, Lykke (2015) *Ethics and Form in Fantasy Literature*. Basingstoke: Palgrave Macmillan.
Guest, Lady Charlotte, trans. (2000) *The Mabinogion*. London: HarperCollins.
Guite, Malcolm (2025) "Malcolm Guite on Enchantment and Imagination," *The Wonder of Tolkien*, 23 January 2025. Available from: https://www.youtube.com/

watch?v=qgM1U7cyqbY&lc=Ugxkw0Amrq52jXwDHBd4AaABAg. Accessed 30 March 2025.
Haars, Stefan (2024) *Get Away*. Big Safari.
Haggard, Piers (1979) *Quatermass*. Euston Films/Thames Television.
Hairston, Andrea (2011) *Redwood and Wildfire*. New York: Tor.
Harding, Stewart (2022) "The Long Walk to Glastonbury—Day 10—Wednesday 23 May 1973." *Facebook*. Available from: https://www.facebook.com/search/top/?q=walking%20to%20Avebury. Accessed 7 September 2022.
Hardy, Robin (1973) *The Wicker Man*. London: British Lion Films.
Hardy, Stephen Paul (2000) "Placiality: The Renewal of the Significance of Place in Modern Cultural Theory." *Brno Studies in English*, vol. 26, no. 1, pp. 85–100. Available from: https://hdl.handle.net/11222.digilib/104271. Accessed 11 April 2025.
Hardy, Thomas (1891) *Tess of the D'Urbervilles: A Pure Woman*. London: James R. Osgood, McIlvaine & Co.
Harrison, M. John (2000) *Viriconium*. London: Gollancz.
Harrison, M. John (2020) *The Sunken Land Begins to Rise Again*. London: Victor Gollancz.
Harrison, Melissa (2021) "Thanks to the Little Grey Men, I Grew Up in Love With a Vision of England I Couldn't Find Anywhere," *The New Statesman*, 8 December 2021. Available from: https://www.newstatesman.com/culture/books/2021/12/the-childrens-books-that-shaped-us-melissa-harrison-on-the-little-grey-men. Accessed 26 February 2024.
Haskell, David (2022) *Sounds Wild and Broken: Sonic Marvels, Evolution's Creativity and the Crisis of Sensory Extinction*. London: Faber and Faber.
Hawthorne, N. (1850) "The Custom House" in *The Scarlet Letter: A Romance*. Available from: http://www.bookwolf.com/Wolf/pdf/NathanielHawthorwne-TheScarletLetterIntroduction.pdf. Accessed 3 April 2018.
Headey, Maria Dahvana (2018) *The Mere Wife*. New York: Farrar, Straus and Giroux.
Heaney, Seamus (2000) *Beowulf*. London: Faber and Faber.
Hergenrader, Trent (2019) *Collaborative Worldbuilding for Writers and Gamers*. London: Bloomsbury Academic.
Hilton, Aoife (2024) "A New Definition for Literary Fantasy: Looking Beyond Publishing Elitism." *BFS Journal*, vol. 25, pp. 130–142. Birmingham: British Fantasy Society.
Hoban, Russell (1980) *Riddley Walker*. London: Jonathan Cape.
Hodgson Burnett, Frances (1911) *The Secret Garden*. London: William Heinemann.
Hoffman, Arthur Sullivant (1923) *Fiction Writers on Fiction: Advice, Opinions and a Statement of Their Own Working Methods by More Than One Hundred Authors*. Indianapolis: IN: Bobbs-Merrill Company.
Holdstock, Robert (1984) *Mythago Wood*. London: Grafton.
Holdstock, Robert (1985) *The Emerald Forest*. London: Penguin.
Holdstock, Robert (1988) *Lavondyss*. London: Grafton.
Holdstock, Robert (2009) *Avilion*. London: Gollancz.
Honko, L. (ed.) (2002) *The Kalevala and the World's Traditional Epics*, vol. 12. Helsinki: Finnish Literature Society.
Hope Hodgson, William (1908) *The House on the Borderland*. London: Chapman and Hall.
Hopkinson, Nalo (2004) *The Salt Roads*. New York: Warner.
Hostetter, Carl F. (ed.) (2021) *The Nature of Middle Earth*. London: HarperCollins.

Hume, Kathryn (2014 [1984]) *Fantasy and Mimesis: Responses to Reality in Western Literature*. London: Routledge.

Huxley, Aldous (1962) *Island*. London: Chatto & Windus.

Ishiguro, Kazuo (2015) *The Buried Giant*. London: Faber and Faber.

ITN (1985) "Battle of the Beanfield, ITN, 1985." Available from: https://www.youtube.com/watch?v=6LHizyCtakw. Accessed 8 September 2022.

Jackson, Rosemary (1998) *Fantasy: The Literature of Subversion*. London: Routledge.

James, Henry (1884 [1888]) "The Art of Fiction," *Longman's Magazine*, p. 4.

James, M. R. (1929) "Some Remarks on Ghost Stories," *The Bookman*, December 1929.

James, Marlon (2019a) *Black Wolf, Red Leopard*. New York: Riverhead Books.

James, Marlon (2019b) "Marlon James: 'You Have to Risk Going Too Far'." *Guardian*, 17 February 2019. Available from: https://www.theguardian.com/books/2019/feb/17/marlon-james-interview-black-leopard-red-wolf. Accessed 9 April 2025.

James, Marlon (2020) *Black Leopard, Red Wolf*. London: Penguin.

James, Marlon (2022) *Moon Witch, Spider King*. London: Penguin.

Jameson, Fredric (1994) *The Seeds of Time*. New York: Columbia University Press.

Jansson, Tove (1945) *Småtrollen och den stora översvämningen/The Moomins and the Great Flood*. Helsinki: Schildts Förlags Ab.

Jansson, Tove (1946) *Kometjakten/Comet in Moominland*. Helsinki: Schildts Förlags Ab.

Jansson, Tove (1957) *Trollvinter/Moominland Winter*. Helsinki: Gebers.

Jansson, Tove (1970) *Moominland in November*. Helsinki: Gebers.

Jansson, Tove (2022 [1972]) *Sommarboken/The Summer Book*. London: Sort of Books.

Jarman, Derek, dir. (1971) "A Journey to Avebury," *BFI Archive*.

Jefferies, Richard (1881) *Wood Magic*. London: Cassell, Petter, Galpin & Co.

Jefferies, Richard (1882) *Bevis: The Story of a Boy*. London: Sampson Low, Marston, Searle, & Rivington.

Jefferies, Richard (1980 [1885]) *After London: Or Wild England*. Oxford: Oxford University Press.

Jemisin, N. K. (2016a) *The Fifth Season*. London: Orbit.

Jemisin, N. K. (2016b) *The Obelisk Gate*. London: Orbit.

Jemisin, N. K. (2017) *The Stone Sky*. London: Orbit.

Jemisin, N. K. (2018) *How Long 'Til Black Future Month?'* London: Orbit.

Jemisin, N. K. (2020) *The City We Became*. New York: Orbit.

Juhren, Marcella (1994) "The Ecology of Middle Earth." *Mythlore: A Journal of J.R.R. Tolkien, C.S. Lewis, Charles Williams, and Mythopoeic Literature*, vol. 20, no. 2, Article 1. Available from: https://dc.swosu.edu/mythlore/vol20/iss2/1.

Jung, Carl (1964) "Introduction" in *Man and His Symbols*. New York: Dell.

Karlin, Daniel (2015, 23 December) *How the Stories Got Their Name: Kipling and the Origins of the 'Just-So' Stories*. Oxford University Press. Accessed 27 October 2016.

Kay, Guy Gavriel (2006 [1985]) *The Summer Tree*. London: Voyager.

Keene, Adrienne (2015, 2 August) "The Benefits of Cultural Sharing Are Usually One-Sided," *The New York Times*. Accessed 14 April 2019.

Keiller, Patrick (1997) *Robinson in Space*. London: BFI.

Kincaid, Paul (2011) *Into the Woods: Robert Holdstock Remembered*. Warton: BSFA.

King, Clive (1963) *Stig of the Dump*. London: Puffin.

Kipling, Rudyard (1894) *The Jungle Book*. London: Macmillan and Co.
Kipling, Rudyard (1902) *Just So Stories*. London: Macmillan and Co.
Kipling, Rudyard (1906) *Puck of Pook's Hill*. London: Macmillan and Co.
Kirk, Robert (2006) *The Secret Commonwealth of Elves, Fauns and Fairies*. New York: New York Review of Books.
Kirk, Tanya, and Matthew Sangster (2023) *Realms of Imagination: Essays From the Wide Worlds of Fantasy*. London: The British Library.
Kolb, David (2022) Sprawling Places. Available from: www.dkolb.org/sprawling-places/generalo/placesto/edwardca.html Accessed 13 October 2022.Krznaric, Roman (2020) *The Good Ancestor: How to Think Long Term in a Short-Term World*. New York: The Experiment.
Kuang, R. F. (2018) *The Poppy War*. London: Harper Voyager.
Landy, J., & Saler, M. (eds) (2009). *The Re-Enchantment of the World: Secular Magic in a Rational Age* (1st ed.). Standord, CA: Stanford University Press. https://doi.org/10.2307/j.ctvr33bqt
Lanham, J. Drew (2022) "My Wild-Like Refuge," *Emergence Magazine: Living With the Unknown*. Available from: https://emergencemagazine.org/essay/my-wild-like-refuge/. Accessed 8 September 2022.
Lavalle, Victor (2016) *The Ballad of Black Tom*. New York: Tor.
Lavalle, Victor (2018) *The Changeling*. Edinburgh: Canongate Press.
Le Guin, Ursula K. (1979) *Language of the Night: Essays on Fantasy and Science Fiction*. New York, NY: Ultramarine Publishing.
Le Guin, Ursula K. (2018) *The Book of Earthsea*. London: Victor Gollancz.
Lee, Fonda (2017) *Jade City*. London: Orbit.
Lee, Fonda (2019) *Jade War*. London: Orbit.
Lee, Fonda (2021) *Jade Legacy*. London: Orbit.
Lee, Laurie (1959) *Cider With Rosie*. London: Hogarth Press.
Lee, Yoon Ha (2016) *Ninefox Gambit*. Oxford: Solaris.
Lefebvre, Henri (1991) *The Production of Space*, vol. 142. Translated by Donald Nicholson-Smith. Oxford: Wiley-Blackwell.
Lethem, Jonathan (2007) "The Ecstasy of Influence: A Plagiarism," *Harpers*, February 2007.
Levy, Michael, and Farah Mendlesohn (2016) *Children's Fantasy Literature: An Introduction*. Cambridge: Cambridge University Press.
Lewis, C. S. (1954) *The Horse and His Boy*. London: Geoffrey Bles.
Lewis, C. S. (1955) *The Magician's Nephew*. London: The Bodley Head.
Lewis, Gwyneth, and Rowan Williams, trans. (2020) *The Book of Taliesin: Poems of Warfare and Praise in an Enchanted Britain*. London: Penguin.
Lewis-Jones, Huw (2018) *The Writer's Map: An Atlas of Imaginary Lands*. London: Thames and Hudson.
Lezard, Nicholas (2008) 'There's Life in the Green Giant Yet'. *Guardian*, 8 March 2008. https://www.theguardian.com/books/2008/mar/08/simonarmitage. Accessed 12 June 2025.
Lindner, Christoph, and Miriam Meissner (eds) (2019) *Routledge Companion to Urban Imaginaries*. London: Routledge.
Lioi, Anthony (2016) *Nerd Ecology: Defending the Earth With Unpopular Culture*. London: Bloomsbury.
Liu, Ken (2015) *The Grace of Kings*. London: Head of Zeus.
Lord, Albert B. (1976) *The Singer of Tales*. New York: Atheneum.
Lowery, David (2021a) "The Green Knight: Screenplay." Available from: https://thescriptlab.com/wp-content/uploads/scripts/36474-THE-GREEN-KNIGHT-by-David-Lowery-.pdf. Accessed 9 April 2025.

Lowery, David, dir. (2021b) *The Green Knight*, A24.
Lynch, David (1977) *Eraserhead*. Libra Films.
Lynch, David (1980) *The Elephant Man*. Paramount Pictures.
Lynch, David (1986) *Blue Velvet*. De Laurentiis Entertainment Group.
Lynch, David (1990–1991) *Twin Peaks*. ABC.
Lynch, David (2001) *Mulholland Drive*. Universal Pictures.
Lynch, David (2017) *Twin Peaks*. Showtime.
Maas, Sarah J. (2015) *A Court of Thorn and Roses*. London: Bloomsbury.
Mabey, Richard (2010) *Weeds: The Story of Outlaw Plants*. London: Profile.
Macfarlane, Robert (2007) *The Wild Places*. London: Granta.
Macfarlane, Robert (2015) *Landmarks*. London: Hamish Hamilton.
Macfarlane, Robert (2025) *Is A River Alive?* London: Hamish Hamilton.
Macfarlane, Robert, Steve Marshall, and Gill Clarke (2021) *Unsettling Landscapes: The Art of the Eerie*. Bristol. Sansom and Company.
Machen, Arthur (1924) *The London Adventure or the Art of Wandering*. London: Martin Secker.
Mackesy, Charles (2019) *The Boy, the Mole, the Fox, and the Horse*. London: Ebury Press.
Maitland, S. (2012) *Gossip from the Forest: The Tangled Roots of Our Forests and Fairytales*. London: Granta.
Manguel, Alberto, and Gianni Guadalupi (1999) *The Dictionary of Imaginary Places*. London: Bloomsbury.
Manwaring, Kevan (2006) *The Bardic Handbook*. Glastonbury: Gothic Image.
Manwaring, Kevan (2008) *Lost Islands: Inventing Avalon, Destroying Eden*. Loughborough: Heart of Albion Press.
Manwaring, Kevan (2014) *Desiring Dragons; Creativity, Imagination, and the Writer's Quest*. Winchester: Compass Books.
Manwaring, Kevan (2018a) *The Knowing: A Fantasy an Epistemological Enquiry into Creative Process, Form, and Genre*. University of Leicester. Thesis. Available from: https://hdl.handle.net/2381/43111.
Manwaring, Kevan (2018b) "Ways Through the Wood: The Rogue Cartographies of Robert Holdstock's Mythago Wood Cycle as a Cognitive Map for Creative Process in Fiction." *Writing in Practice*, vol. 4. NAWE. Available from: https://www.nawe.co.uk/DB/wip-editions/articles/ways-through-the-wood-the-rogue-cartographies-of-robert-holdstocks-mythago-wood-cycle-as-a-cognitive-map-for-creative-process-in-fiction.html. Accessed 6 September 2024.
Manwaring, Kevan (2020a) "Walking Between Worlds: In Defence of Experiential Research." *Writing in Practice*, vol. 6. Available from: https://issuu.com/publications-nawe.co/docs/wip_2020/144.
Manwaring, Kevan (2020b) "Finding the Line: A Triangulation Between Walking, Multimodality, and Embodied Poetics." *Axon: Poetry: Small Leaps, Giant Steps*, vol. 10, no. 1. Available from: https://www.axonjournal.com.au/issue-vol-10-no-1-may-2020/finding-line.
Manwaring, Kevan (2020c) "Lines of Desire: The Phenomenology of Long-Hand Writing in Creative Praxis" in *New Writing: The International Journal for the Practice and Theory of Creative Writing*, vol. 17, no. 2. Available from: https://www.tandfonline.com/doi/abs/10.1080/14790726.2019.1586958.
Manwaring, Kevan (2021) "Performing Kirk: A Search for Authenticity in the Dramatization of the Life of the 'Fairy Minister', Reverend Robert Kirk." *Revenant*, vol. 6, pp. 67–90. Available from: https://www.revenantjournal.com/

contents/performing-kirk-a-search-for-authenticity-in-the-dramatisation-of-the-life-of-the-fairy-minister-reverend-robert-kirk/. Accessed 30 March 2025.

Manwaring, Kevan (2024a) "Fairy and Liminality in the Scottish Lowlands." *BFS Journal*, Winter, vol. 25, pp. 119–129. British Fantasy Society.

Manwaring, Kevan (2024b) *Writing Ecofiction: Navigating the Challenges of Environmental Narratives*. London: Palgrave Macmillan.

Marco, Subcomadante (2001) From *Subcomandante Marcos: Zapatista Stories*. Translated by Dinah Livingstone. London: Katabasis.

Markley, Stephen (2023) *The Deluge*. London: Simon & Schuster.

Masefield, John (1927) *The Midnight Folk*. London: William Heinemann.

Masefield, John (1935) *The Box of Delights, or When the Wolves Were Running*. London: William Heinemann.

Matheson, Richard (1999 [1954]) *I Am Legend*. SF Masterworks. London: Gollancz.

McCarthy, Cormac (2006) *The Road*. New York: Alfred A. Knopf.

McGilchrist, Iain (2009) *The Master and His Emissary: The Divided Brain and the Making of the Western World*. Yale University Press.

McKenzie, Alan, and John Ridgeway (2017 [1988–1995]) *Summer Magic: The Complete Journal of Luke Kirby*, vol. 1. Oxford: Rebellion Publishing.

McKillip, Patricia (2003) *Ombria in Shadow*. London: Gollancz.

Meadows, Donella (2017 [2004]) *Thinking in Systems: A Primer* (originally published in Timeline Magazine, no. 74, March 2004). London: Chelsea Green Publishing. Available from: https://donellameadows.org/dancing-with-systems/. Accessed 20 December 2023.

Mellon, Nancy (1988) *The Art of Storytelling*. Shaftesbury: Dorset.

Mendlesohn, Farah (2008) *Rhetorics of Fantasy*. Middletown, CT: Wesleyan University Press.

Merleau-Ponty, Maurice (1962) *Phenomenology of Perception*. Translated by Colin Smith. London: Routledge & Kegan Paul.

Merlin (2008–2012) BBC TV Series.

Miéville, China (2009) *The City and the City*. London: Macmillan.

Miéville, China (2018) "The Limits of Utopia," *Climate and Capitalism*. Available from: https://climateandcapitalism.com/2018/03/02/china-mieville-the-limits-of-utopia/. Accessed 12 August 2021.

Miller, George (1979) *Mad Max*. Kennedy Miller Productions.

Miller, John (2020) *Weird Woods: Tales From the Haunted Forests of Britain. Tales of the Weird*. London: The British Library.

Miller, Walter M., Jr. (1959) *A Canticle for Leibowitz*. Philadelphia, PE: J.B. Lippincott & Co.

Mirrlees, Hope (1923) *Lud-in-the-Mist*. London: Collins.

Mitchell, Ben (2013) "So Bright, So Beautiful . . . Precious! Cursed Ring Thought to Have Inspired JRR Tolkien on Display," *The Independent*, 2 April 2013. Available from: https://www.independent.co.uk/arts-entertainment/books/news/so-bright-so-beautiful-precious-cursed-ring-thought-to-have-inspired-jrr-tolkien-on-display–8557229.html. Accessed 7 September 2022.

Mitchell, David (2004) *Cloud Atlas*. London: Sceptre.

Miyazaki, Hayao (1997) *Princess Mononoke*. Tokyo: Studio Ghibli.

Miyazaki, Hayao (2004) *Howl's Moving Castle*. Studio Ghibli.

Miyazaki, Hayao (2010) *Arrietty*. Studio Ghibli.

Mohamed, Premee (2020) *Beneath the Rising*. Oxford: Solaris.

Mohamed, Premee (2021) *The Broken Darkness*. Oxford: Solaris.
Mohamed, Premee (2022) *The Void Ascendant*. Oxford: Solaris.
Moorcock, Michael (1968) *The Final Programme*. New York: Avon Books.
Moorcock, Michael (1971) *A Cure for Cancer*. London: Allison & Busby.
Moorcock, Michael (1977) *The Condition of Muzak*. London: Allison & Busby.
Moorcock, Michael (1979) *The English Assassin: A Romance of Entropy*. London: Allison & Busby.
Moore, Alan (2000) *The Highbury Working: A Beat Séance*. Marietta, GE: Top Shelf Productions.
Moore, Alan (2024) *The Great When*. London: Bloomsbury.
Moore, Alan, and Kevin O'Neil (1999–2019) *The League of Extraordinary Gentlemen*. Marietta, GE: Top Shelf Productions.
Moore, Simon (2000) *The 10th Kingdom*. Studio City, CA: Hallmark Entertainment.
Moreno-Garcia, Silvia (2020) *Mexican Gothic*. London: Jo Fletcher Books.
Morgenstern, Erin (2011) *The Night Circus*. London: Vintage.
Myrick, Daniel, dir. (1999) *The Blair Witch Project*. Haxan Films.
Nanson, Anthony (2021) *Storytelling and Ecology; Empathy, Enchantment and Emergence in the Use of Oral Narratives*. London: Bloomsbury Academic.
Nanson, John (2023) *Ecobardic Desire: Storytelling and Storywriting as Catalysts of Transformation in People's Relationship with the More-than-Human World*. University of Gloucestershire. PhD Thesis. Available from: https://doi.org/10.46289/7RK3P8X4.
National Trust (2022) "New Discovery at Alderley Edge in Cheshire Offers a Glimpse into a Mine in Production During the Napoleonic Wars, With Personal Objects, Equipment and Inscriptions the Workers Left Behind," *National Trust*, 12 July 2022. Available from: https://www.nationaltrust.org.uk/press-release/rare-time-capsule-cobalt-mine-abandoned-over-200-years-ago-is-discovered-in-cheshire. Accessed 8 September 2022.
Ng, Jeanette (2017) *Under the Pendulum Sun*. Nottingham: Angry Robot.
Nicholls, Stan (1993) *Wordsmiths of Wonder*. London: Orbit.
Norton, Mary (1952) *The Borrowers*. London: J.M. Dent.
Novik, Naomi (2015) *Uprooted*. London: Macmillan.
Ogbunu, C. Brandon (2023) "Nnedi Okorafor, a Pioneer of Africanfuturism, Doesn't Want Her Work Put in a Box," *Andscape*. Available from: https://andscape.com/features/nnedi-okorafor-a-pioneer-of-africanfuturism-doesnt-want-her-work-put-in-a-box/. Accessed 9 April 2025.
Okorafor, Nnedi (2010) *Who Fears Death*. New York: DAW.
Okorafor, Nnedi (2014) *Lagoon*. London: Hodderscape.
Okorafor, Nnedi (2015) *Binti*. London: Tor.
Okorafor, Nnedi (2016) "Nnedi Okorafor: Writing Oneself into Existence," *Mail & Guardian*, 16 September 2016. Available from: https://mg.co.za/article/2016-09-16-00-nnedi-okorafor-writing-oneself-into-existence/. Accessed 13 October 2022.
Olusoga, David (2016) *Black and British: A Forgotten History*. London: Pan Macmillan.
Opie, James, and Iona Opie (1951) *The Oxford Dictionary of Nursery Rhymes*. Oxford: Clarendon Press.
Orlowski-Yang, Jeff (2020) *The Social Dilemma*. Exposure Labs.
Orme, Stuart (1980) *Edge of the Ceiling*. Granada TV.
Overall, Sonia (2021) *Heavy Time*. London: Penned in the Margins.

Owens, Susan (2021) *Spirit of Place: Artists, Writers & the British Landscape.* Thames & Hudson.
Oxford Reference (2022) "Oxford Reference." Available from: https://www.oxfordreference.com/view/10.1093/oi/authority.20110803100114463. Accessed 9 September 2022.
Palmer-Patel, Chuckie (2020) "The Shape of Fantasy: Investigating the Structure of American Heroic Epic Fantasy" in *Routledge Research in American Literature and Culture*. New York, NY: Routledge.
Papadimitriou, Nick (2013) *Scarp*. London: Sceptre.
Parker, Elizabeth (2020) *The Forest and the EcoGothic: The Deep Dark Woods in the Popular Imagination*. London: Palgrave Macmillan.
Parker-Chan, Shelley (2021) *She Who Became the Sun*. London: Pan.
Parker-Chan, Shelley (2023) *He Who Drowned the World*. London: Pan.
Peake, Mervyn (1992) *Gormenghast*. London: Mandarin.
Phelan, James (2005) *Living to Tell About It: A Rhetoric and Ethics of Character Narrative*. Columbus: Ohio State University Press.
Philip, Neil (1981) *A Fine Anger: A Critical Introduction to the Work of Alan Garner*. London: Collins.
Poe, E. A. (1842) "The Importance of the Single Effect in a Prose Tale" in *Graham's Magazine*. Philadelphia, PA: G.R. Graham.
Poe, E. A. (1846) "The Philosophy of Composition." Available from: http://short-storyamerica.com/pdf_classics/poe_philosophy_of_composition.pdf. Accessed 4 April 2018.
Poe, Edgar Allan (1840) "The Man in the Crowd" in *Burton's Gentleman's Magazine*. Philadelphia.
Pratchett, Terry (1983) *The Colour of Magic*. Bucks: Colin Smythe.
Propp, Vladimir (1968 [1928]) *The Morphology of the Folk Tales*. 2nd revised edn. Texas: University of Texas Press.
Pullman, Philip (2001) *His Dark Materials*. London: Scholastic.
Pullman, Philip (2017) *The Book of Dust, Vol. 1: La Belle Sauvage*. Oxford: David Fickling Books.
"Puzzlewood Through the Lens of Thomas Duke." Available from: https://www.puzzlewood.net/index.php/component/k2/item/29-puzzlewood-through-the-lens-of-thomas-duke. Accessed 7 September 2022.
Qing, Li (2022) "Effects of Forest Environment (Shinrin-yoku/Forest Bathing) on Health Promotion and Disease Prevention—The Establishment of 'Forest Medicine'." *Environmental Health and Preventive Medicine*, vol. 27, p. 43. Available from: https://doi.org/10.1265/ehpm.22–00160.
"Quantum Entanglement." Available from: https://en.wikipedia.org/wiki/Quantum_entanglement. Accessed 13 October 2022.
Ratchford, Fannie E. (1941) *The Brontës' Web of Childhood*. New York: Columbia University Press.
Rayner, Alan (2006) "Alan Rayner on Inclusionality, Boundaries and Space. Recorded by Jack Whitehead." *Youtube*. Available from: https://youtu.be/yVa-7FUIA3W8. Accessed 12 July 2023.
Rayner, Alan (2010) "Inclusionality and Sustainability-Attuning With the Currency of Natural Energy Flow and How This Contrasts With Abstract Economic Rationality." *Environmental Economics*, vol. 1, pp. 98–108.
Read, Rupert (2017) "THRUTOPIA: Why Neither Dystopias Nor Utopias Are Enough to Get Us Through the Climate Crisis, and How a 'Thrutopia' Could Be,"

Huffington Post, 6 November 2017. Available from: https://www.huffingtonpost.co.uk/rupert-read/thrutopia-why-neither-dys_b_18372090.html. Accessed 11 April 2025.
Reed, Carol (1949) *The Third Man*. London Films.
Reyes, Rowena Capulong (2016, March) "Public Space as Contested Space: The Battle Over the Use, Meaning and Function of Public Space." *International Journal of Social Science and Humanity*, vol. 6, no. 3.
Rice, Anne (1976) *An Interview With the Vampire*. New York: Alfred A. Knopf.
Rice, Anne (1990) *The Witching Hour*. New York: Alfred A. Knopf.
Roanhorse, Rebecca (2020) *Black Sun*. Oxford: Solaris.
Robinson, Joanna (2021, July) "The Green Knight's Ending Explained," *Vanity Fair*. Available from: https://www.vanityfair.com/hollywood/2021/07/green-knight-ending-explained-does-he-die-gawain-dev-patel. Accessed 8 September 2022.
Robinson, Kim Stanley (2021) *The Ministry for the Future*. New York: Orbit.
Routh, H. V. (1913) "The Future of Comparative Literature." *Modern Language Review*, vol. 8, no. 1, pp. 1–14.
Rowling, J. K. (1997) *Harry Potter and the Philosopher's Stone*. London: Bloomsbury.
Ruff, Matt (2016) *Lovecraft Country*. New York: HarperCollins.
Russell, George William (1918) *The Candle of Vision*. London: Macmillan.
Said, S. F. (2022) *Tyger*. Oxford. David Fickling.
Saint-Exupery, Antoine de (2018 [1945]) *The Little Prince*. Stansted: Wordsworth Classics.
Samatar, Sofia (2014) *A Stranger in Olondria*. Easthampton, MA: Small Beer Press.
Sanders, Rupert, dir. (2012) *Snow White and the Huntsman*. Roth Films.
Sanders, Rupert, dir. (2016) *The Huntsman: Winter's War*. Roth Films.
Sanderson, David (2019, 3 December) "Booker Winner Bernardine Evaristo Writes Off 'Cultural Appropriation'," *The Times*. Accessed 13 October 2022.
Sandner, D. (2011) *Critical Discourses of the Fantastic, 1712–1831* (1st ed.). London: Routledge. https://doi.org/10.4324/9781315574974
Sangster, Matthew (2023) *An Introduction to Fantasy*. Cambridge: Cambridge University Press.
Sapochnik, Miguel (2021) *Finch*. Apple TV.
Saunders, Erwin (2022) "First Sighting of a Pixie." Available from: https://www.youtube.com/watch?v=3GsyM6–5-QM&t=384s. Accessed 8 September 2022.
Schama, Simon (1995) *Landscape and Memory*. New York: Alfred A. Knopf.
Scott, Manda (2023) "We're Clearly in the Midst of a *#metacrisis*. Linear Thinking Got Us into This Mess—How Can We Engage All the Tools of Modern Technology to Help Us Think More Freely, Creatively and Systemically in Service to a More Flexible Future?," *Facebook*, 22 February 2023. Available from: https://www.facebook.com/MandaScottAuthor/posts/were-clearly-in-the-midst-of-a-metacrisis-linear-thinking-got-us-into-this-mess-/601025515173706/. Accessed 9 April 2025.
Scott, Manda (n.d.) "A New Project for a New World," *The Blog*. Available from: https://mandascott.co.uk/a-new-project-for-a-new-world/. Accessed 9 April 2025.
Scott, Peter Graham, dir. (1977) *The Children of the Stones*. HTV.
Scurr, R. (2015) *John Aubrey: My Own Life*. London: Chatto & Windus.
Sendark, Maurice (2000 [1963]) *Where the Wild Things Are*. London: Red Fox.

Sethi, Anita (2021) *I Belong Here: A Journey Along the Backbone of Britain.* Bloomsbury Publishing.
Shakespeare, William (1991[1604]) *Hamlet.* Oxford Text Archive Core Collection.
Shaw, Martin (2023) *Courting the Wild Twin.* London: Chelsea Green Publishing.
Shawl, Nisi, and Cynthia Ward (2005) "Writing the Other: A Practical Approach" in *Conversation Pieces*, vol. 8. Aqueduct Press.
Sheldrake, Merlin (2020) *Entangled Life: How Fungi Make Our Worlds, Change Our Minds, and Shape Our Futures.* London: Bodley Head.
Sheldrake, Merlin (2022) *Planet Earth III.* London: BBC. Available from: http://www.bbc.com/earth/story/20141111-plants-have-a-hidden-internet. Accessed 13 October 2022; Accessed 11 April 2025.
Shen, Darren (2008) "What is 'Implied Author'?" *Journal of Peking University (Philosophy and Social Sciences)*, p. 2.
Shepherd, Nan (2011 [1977]) *The Living Mountain.* Edinburgh: Canongate.
Shriver, Lionel (2016) "Lionel Shriver's Full Speech: 'I Hope the Concept of Cultural Appropriation is a Passing Fad'," *Guardian*, 13 September 2016. Available from: https://www.theguardian.com/commentisfree/2016/sep/13/lionel-shrivers-full-speech-i-hope-the-concept-of-cultural-appropriation-is-a-passing-fad. Accessed 13 October 2022.
Simonson, Martin (ed.) (2015) *Representations of Nature in Middle-Earth.* Zurich: Walking Tree Publishers.
Sinclair, Iain (1997) *Lights Out for the Territory: 9 Excursions in the Secret History of London.* London: Granta.
Sinclair, Iain (2003) *London Orbital.* London: Penguin.
Singer, Bryan, dir. (2013) *Jack the Giant Slayer.* New Line Cinema.
Sleigh, Bernard, and Sidgwick & Jackson (1920) *An Anciente Mappe of Fairyland: Newly Discovered and Set Forth: Imaginary Locality.* London. Retrieved from the Library of Congress, Geography and Map Division.
Smith, Phil (2010) *Mythogeography: A Guide to Walking Sideways.* Charmouth: Triachy Press.
Smith, David Woodruff (2013) '*Phenomenology*'. Stanford, CA: The Stanford Encyclopedia of Philosophy. Available from https://plato.stanford.edu/entries/phenomenology/. Accessed 12 June 2025.
Smith Spark, Anna (2025) "In Conversation With: Anna Smith Spark." Interviewed by Kevan Manwaring. *BFS Journal*, vol. 26. Birmingham: British Fantasy Society.
Snyder, Gary (1990) *The Practice of the Wild.* Berkeley: Counterpoint.
Solnit, R. (2013) "A Thousand Stories in the Naked City', Part of Creative Time Summit: 'Art, Place, & Dislocation in the 21st Century City." Available from: https://www.youtube.com/watch?v=L9kWI2skHdw. Accessed 13 October 2022.
Solnit, Rebecca (2001) *Wanderlust: A History of Walking.* London: Penguin.
Sondheim, Stephen (1986) *Into the Woods.* New York: Broadway.
Spivak, Gayatri (2017) "Cited in Majumder, Auritro, 'Gayatri Spivak, Planetarity and the Labor of Imagining Internationalism." *Mediations: Journal of the Marxist Literary Group*, vol. 30, no. 2. *Post-Humanisms Reconsidered.* Available from: https://www.mediationsjournal.org/articles/planetarity. Accessed 13 October 2022.
Spufford, Francis (2002) *The Child That Books Built: A Memoir of Childhood and Reading.* London: Faber and Faber.

"Stanislavski 'Method'." Available from: https://www.britannica.com/art/Stanislavsky-system. Accessed 13 October 2022.
Steig, William (1990) *Shrek!* New York: Farrar, Straus and Giroux.
Steinberg, Avi (2011) "The Grand Map." *The Paris Review*, 5 October. https://www.theparisreview.org/blog/2011/10/05/the-grand-map/. Accessed 11 April 2025.
Steinberg, Saul (1976) 'View of the World From 9th Avenue & Steinbergian Cartography." Available from: https://saulsteinbergfoundation.org/essay/view-of-the-world-from-9th-avenue/. Accessed 13 October 2022.
"Stepping Through Film." Available from: https://www.instagram.com/steppingthroughfilm/. Accessed 7 September 2022.
Stevenson, Robert Louis (1886) *The Strange Case of Dr Jekyll and Mr Hyde*. London: Longmans, Green, and Co.
Stevenson, Robert Louis (1888) "A Chapter on Dreams," *Scribner's Magazine*. New York: Charles Scribner & Sons, pp. 122–127.
Stewart, George (2015 [1949]) *Earth Abides*. London: Gollancz.
Stibbe, Arran (2021) *Ecolinguistics: Language, Ecology and the Stories We Live By*. Abingdon: Routledge.
Stroud, Allen (2023) *Historical Dictionary of Fantasy Literature*. London: Rowman & Littlefield.
Strugatsky, Arkady, and Boris (1972) *Roadside Picnic*. London: Macmillan.
Suri, Tasha (2021) *The Jasmine Throne*. London: Orbit.
Swift, Jonathan (1726) *Gulliver's Travels*. London: Benjamin Motte.
Tarkovsky, Andrei (1979) *Stalker*. Moscow: Mosfilm.
Taylor, Charles (2007) *A Secular Age*. Cambridge, MA: Harvard University Press.
Teffau, Lauren C. (2024) *The Hunger With No Name*. Tampa, FL: University of Tampa Press.
Tesh, Emily (2019) *Silver in the Wood*. London: Tor.
Thomas, Ebony Elizabeth (2019) *The Dark Fantastic*. New York University Press.
Thompson, Tade (2016) *Rosewater*. London: Little, Brown.
Thompson, Tade (2019) *The Rosewater Insurrection*. London: Little, Brown.
Thompson, Tade (2019) *The Rosewater Redemption*. London: Little, Brown.
Thorkelsson, Svanur (2020) *Beowulf*. London: Audible Studios.
Thorpe, Adam (2014) *On Silbury Hill*. Beaminster: Little Toller.
Todd, Barbara Euphan (1936) *Worzel Gummidge*. London: Burns Oates & Washbourne Ltd.
Todorov, Vladimir (1975) *The Fantastic: A Structural Approach to a Literary Genre*. Ithaca, NY: Cornell University Press.
Toffler, Alvin (1980) *The Third Wave*. New York: William Morrow.
Tolkien, J. R. R. (1932) "'The Name Nodens', Appendix to 'Report on the Excavation of the Prehistoric, Roman and Post-Roman Site in Lydney Park, Gloucestershire," Reports of the Research Committee of the Society of Antiquaries of London, 1932; also in *Tolkien Studies: An Annual Scholarly Review*, vol. 4, 2007.
Tolkien, J. R. R. (1964) "Interviewed by Irene Slade for Reluctant Olympians," *Broadcast*, 26 November 1964. London: BBC.
Tolkien, J. R. R. (1991a) *The Lord of the Rings*. London, England: HarperCollins.
Tolkien, J. R. R. (1991b) *The Silmarillion*. London, England: HarperCollins.
Tolkien, J. R. R. (1997) *The Monsters and the Critics and Other Essays*. London: HarperCollins.
Tolkien, J. R. R. (1998) *Tales From the Perilous Realm*. London: HarperCollins.
Tolkien, J. R. R. (2002) *The Book of Lost Tales*. London: HarperCollins.

Tolkien, J. R. R. (2008) *Tales From the Perilous Realm*. New York, NY: Houghton Mifflin Harcourt.
Tolkien, J. R. R. (2012) *The Hobbit*. New York, NY: Houghton Mifflin Harcourt.
Tolkien, J. R. R., and Christopher Tolkien (ed.) (2020) *Sir Gawain and the Green Knight, with Pearl and Sir Orfeo*. London: HarperCollins.
"Tolkien Trail—Birmingham City Council." Available from: https://www.birmingham.gov.uk/downloads/file/9230/birmingham_tolkien_trail.
Tooze, Adam (2022) "Welcome to a World of the Polycrisis," *Financial Times*, 28 October 2022. Available from: https://www.ft.com/content/498398e7–11b1–494b-9cd3–6d669dc3de33. Accessed 16 March 2023.
Turchi, Peter (2004) *Maps of the Imagination: The Writer as Cartographer.* San Antonio, TS: Trinity University Press.
Tuttle, Lisa (2012) *The Silver Bough*. London: Jo Fletcher Books.
Twain, Mark (1884) *Adventures of Huckleberry Finn*. London: Chatto & Windus.
VanderMeer, Jeff (2013) *Wonderbook: The Illustrated Guide to Creating Imagination Fiction*. New York, NY: Abrams Image.
VanderMeer, Jeff (2014) *Area X: The Southern Reach Trilogy*. London: 4th Estate.
VanderMeer, Jeff (2024) *Absolution*. London: 4th Estate.
Vaughan-Lee, E., and A. Loften (2019) "'Counter-Mapping' by Adam Loften & Emmanuel Vaughan-Lee," *Emergence*. Available from: https://emergencemagazine.org/story/counter-mapping/. Accessed 13 October 2022.
Villanueva-Romero, D., Kerslake, L., & Flys-Junquera, C. (Eds.). (09 Dec. 2021). Imaginative Ecologies. Leiden, The Netherlands: Brill. https://doi.org/10.1163/9789004501270
Waal, Kit de (2018) "Don't Dip Your Pen in Someone Else's Blood: Writers and 'The Other'," *The Irish Times*, 30 June 2018. Available from: https://www.irishtimes.com/culture/books/don-t-dip-your-pen-in-someone-else-s-blood-writers-and-the-other–1.3533819. Accessed 13 October 2022.
Wall Kimmerer, Robin (2020 [2013]) *Braiding Sweetgrass: Indigenous Wisdom, Scientific Knowledge and the Teachings of Plants*. London: Penguin.
Walton, Alexandra (2022) "The Archival Trail: Paul Nash the War Artist," *Tate Britain*. Available from: https://www.tate.org.uk/art/artists/paul-nash-1690/paul-nash-war-artist. Accessed 8 September 2022.
Warner, Marina (1995) *From the Beast to the Blonde: On Fairy Tales and Their Tellers*. London: Vintage.
Warner, Sylvia Townsend (1926) *Lolly Willowes*. London: Chatto & Windus.
Warner, Sylvia Townsend (1977) *Kingdoms of Elfin*. London: Chatto & Windus.
Wasserman, E. R. (1950) "The Inherent Values of Eighteenth-Century Personification." *Publications of the Modern Language Association of America*, vol. 65, no. 4, pp. 435–463.
Watkins, Alfred (1974 [1925]) *The Old Straight Track*. London: Abacus.
Watson, Emma (2022) "Emma Watson Interviews Rebecca Solnit," *Off the Shelf*, 12 December 2019. Available from: https://www.youtube.com/watch?v=yt4vOWBf6WA. Accessed 13 October 2022.
Welfare, Simon, John Fairley, and Arthur Charles Clarke (1980) *Arthur C. Clarke's Mysterious World*. Galahad Books.
Well, H. G. (1895) *The Time Machine: An Invention*. London: William Heinemann.
Well, H. G. (1898) *The War of the Worlds*. London: William Heinemann.
Westling, L. 'Merleau-Ponty and the Eco-Literary Imaginary,' in Handbook to Ecocriticism and Cultural Ecology, edited by Hubert Zapf.. Gruyter Verlag, 2016.

White, T. H. (1958) *The Once and Future King.* London: Putnam.
Whitehead, Colson (2016) *The Underground Railroad.* New York: Knopf Doubleday.
Whitehead, Harry (2013) "Nomadic Emergence: Creative Writing Theory and Practice-Led Research" in Harper, Graeme (ed.) *New Ideas in the Writing Arts.* Newcastle: Cambridge Scholars Publishing, p. 101.
Whybrow, Helen, and Robin Wall Kimmerer (2017) "Robin Wall Kimmerer on the Language of Animacy," *Orion Magazine*, March/April. Published online 17 July 2017. Available from: https://orionmagazine.org/article/robin-wall-kimmerer-language-animacy/. Accessed 6 September 2024.
Wilczek, Frank. "Entanglement Made Simple," *Quanta Magazine.* Available from: https://www.quantamagazine.org/entanglement-made-simple-20160428. Accessed 13 October 2022.
Williams, Charles (1930) *War in Heaven.* London: Victor Gollancz.
Wilson, G. Willow (2012) *Alif the Unseen.* London: Corvus.
Windling, Terri, and Delia Sherman, eds. (1998 [1986]) *The Essential Bordertown.* New York: St Martin's Press.
Winterson, Jeanette (1992) *Oranges Are Not the Only Fruit.* London: Vintage.
Winton, Tim (2024) *Juice.* London: Picador.
Wolfman, Marv, and Gene Colan (1982) *Night Force.* Burbank, CA: DC Comics.
World Bank (2024) "Urban Development." Available from: https://www.worldbank.org/en/topic/urbandevelopment/overview#:~:text=Today%2C%20some%2056%25%20of%20the,people%20will%20live%20in%20cities. Accessed 9 April 2025.
Wyndham, John (1951) *The Day of the Triffids.* London: Michael Joseph.
Wyndham, John (1957) *The Midwich Cuckoos.* London: Michael Joseph.
Wynne-Jones, Diana (1986) *Howl's Moving Castle.* London: HarperCollins.
Wynne-Jones, Diana (2004) *The Tough Guide to Fantasy Land.* London: Gollancz.
Yunkaporta, Tyson (2024) "Deep Time Diligence: An Interview With Tyson Yunkaporta," *Emergence Magazine*, 19 February 2024. Available from: https://emergencemagazine.org/conversation/deep-time-diligence/. Accessed 1 April 2025.
Zemeckis, Robert (2007) *Beowulf.* Paramount Pictures.

Index

#MeToo 125
28 Days Later 120
1980s 102
2000 AD 121

Aarne-Thompson-Uther Index 52, 53
Abercrombie, Joe 93
Absolute Space 169, 170
Absolution 110
Acceptance 110
Ackroyd, Peter 107, 156
activistic 110, 149, 192
Adams, Richard 78
Addison, Joseph 10
Adventures of Baron Munchausen, The 96, 144
AE, George William Russell 13
affect 153, 173, 175, 176, 179, 188, 190, 191
Africanfuturism 149
Afrofuturism 149, 157
After London: or Wild England 76, 105, 121, 122, 142
Age of Reason 11, 28, 87, 92
Alain-Fournier 58
Albion 104
Alderley Edge 134, 158, 186–190
Aleph, The 99
Alice's Adventures in Wonderland 60
American Gods 155, 168
An ancient mappe of Fairyland 53
Anatomy of Melancholy, The 3
Ancient Echoes 139
Angria 56
animals 26, 31, 49, 51, 53, 61, 65, 75, 77–79, 84, 105, 118, 141
An Interview with a Vampire 115
Ankh-Morpork 154
Annihilation 110
Annwn 46, 49, 83, 108, 188
Anthropocene 20, 21, 89, 144, 150, 151, 196
Apostle 127
Aquae Sulis 183
Archetypes 58
Area of Outstanding Natural Beauty 174
Area X 110
Aristotle 80
Arnold, Matthew 7
Arrietty 114
Art of the Novel, The 163
Arthur C. Clarke's Mysterious World 178, 179
Arthuriad 84, 132, 168
As You Like It 4–6
assemblage 165, 169
Atlantean 57
Attebery, Brian 9, 20, 25
Authority 110
Autocartography 154, 162–165
Avalon 46, 90, 168
Avatar 194, 195
Avebury 176–180
Avilion 99, 128–130
Ayesha 91

Babel, or The Necessity of Violence 124, 146–148
Bach, Richard 78, 79
Ballard, J.G. 121
Barker, Clive 115
Barker, R.J. 158
Barleycorn, John 128

Barnett, David 135, 136
Barrie, J.M. 61, 65, 66, 69, 94, 174
bastard countryside 107
Bath, city of 183, 186
BB, (D.J. Watkins-Pitchford) 62–65
Beagle, Peter S. 73, 78, 94, 95
Beneath the Rising 150, 172
Beowulf 8, 28, 29, 33–36, 133, 167
Beresford, Elizabeth 114
Beuke, Lauren 141
Bevis 71
Bey, Hakim 156, 179
Bible 194
Big Rock Candy Mountain 89
Bihemispheric brain 5
biosemiotics 5, 18, 21, 24, 25
Biosemiotic Theory of Fantasy 25, 198–200
BIPOC 115, 125, 148, 149, 172
Birmingham 73, 184–186
Black Fantastic (exhibition) 150
Black Leopard, Red Wolf 162, 172
Black Lives Matter 125, 148, 167
Black Panther 149
Blackwood, Algernon 15
Blair Witch Project, The 189
Blake, William 104, 105
Bloemfontein 184
Boneland 134, 187–189
Book of Dust, The 76, 117, 141
Book of Earthsea, The 8, 70, 73, 80, 158, 193
Books of Magic, The 115
Bordertown 110
Borges, Jorge Luis 155
Borrowers, The 114
Boudica 168
Box of Delights, or When the Wolves Were Running, The 75–76
A Boy and His Dog 120
Boy, The Mole, The Fox, and The Horse, The 78
Branwell 56, 57, 146
Brazil 96, 102
Briggflatts 135
British Empire 118, 147
Broken Earth trilogy 124, 150, 161
Brontës *Web of Childhood, The* 56, 57
Bronze Age 177, 178, 186
Brotherhood of English Ruralists 177
Bulgakov, Mikhail 109, 110
Bull, Emma 96, 102
Buried Giant, The 8, 162
Butler, Octavia 148

Cabell, James Branch 89
Caldecott, Moyra 178
Caliban 12, 85, 86
Candle of Vision, The 14
Canterbury Tales, The 101
A Canticle for Leibowitz 120
capitalism 140, 167, 194, 197
Carnival Row 115
Carroll, Lewis 142, 147, 154, 155
Carson, Ann 78
cartography 152–173
Celtic mythology 47, 49, 112, 128
Celtic Underworld 83
Ceridwen 47
Chabon, Michael 14
Charlie and the Chocolate Factory 89
Chester, Nicol 78
Chestertonian 105
Child That Books Built, The 58
Children of Lir 50, 83
Children of the Stones 179
Children's Fantasy 52–79
Children's writing 52–79
Christian 1, 8, 34–37, 92, 143
Chronicles of Thomas Covenant the Unbeliever, The 82, 87
chrono-abyss 50, 83
Circe 168
Circle, The 182
City and The City, The 117
City Born Great, The 118
City We Became, The 118
Clarke, Susanna 142–143, 147
climate change 18, 19, 21, 25, 26, 30, 75, 78, 113, 122, 124, 127, 161, 196
Climate Chaos 141, 195
Climate Crisis 27, 29, 167, 192
Climate Emergency 124, 192
Climate Fiction 122
Climate Strikes 71
Cloud Atlas 132
Cloud Cuckoo Land 124, 143, 144
Clute, John 7–9, 21, 28, 50, 79–81, 83, 99, 111, 114, 156
coastal cities 113
Cockayne, Land of 89

colonialism 146, 148, 157
Comedy of Fantasy, The 22
Comet in Moominland 74
comic books 74, 102, 104, 112, 120, 127, 136, 149, 168
commercial fantasy 86, 93, 200
commodified fantasy 80–82, 93
Conquest of Gaul 125, 159
contested spaces 165
Cooper, Susan 76–77, 115
Cotswolds 184
Cottage of Lost Play 54, 55, 185, 199
counterculture 93
counter-mapping 165–168
A Court of Thorns and Roses 158
Cresswell, Tim 153, 156, 161
Crook, Mackenzie 78
Crowman, The 78
cursed ring 183–184

Daevabad 150, 158, 172
Damaya 71
Dark Fantastic 148–151
Dark Is Rising Sequence, The 8, 76, 77, 115
Dark Other 160
Debord, Guy 103, 156, 181
Delaney, Samuel R. 148
De Lint, Charles 72, 96, 102, 112, 113
Dery, Mark 149, 157
Destinarianism 71, 86
De Waal, Kit 159
Diagon Alley 115
Dickinson, Emily 7
Die Entzauberung der Welt 7
Dirk Gently's Holistic Detective Agency 115
Discworld 152
disenchantment 6, 7–9, 81, 87, 88, 92–98, 119
diversification 123, 125, 148
Doctor Who 75, 102, 115, 178, 180
Doerr, Anthony 143, 144
Door of Durin 184
doughnut economics 113
Dover Beach 7, 92
Dunsany, Lord 29, 61
Dust Bowl 89

Earth Abides 121
Earthsea 8, 26, 70, 73, 80, 158, 193
Echtrai 70, 90
Ecofiction 113, 192
Ecolinguistic/s 21, 22, 87, 200
Ecoliteracy 20, 47, 77
Ecological Imaginary 7, 11, 16–20, 24, 29, 40, 47, 54, 62, 65, 71, 73, 105, 106, 110, 112, 128, 130, 134, 139, 142, 148, 150, 199, 200
ecology 20, 24, 63, 121, 158, 162
Ecopaganism 125–138
Ecosophic 21–23
ecosystem 4, 17, 20, 22, 24, 37, 54, 74, 80, 143, 153, 182, 193, 194, 199
Eden 13, 28, 56, 69
Edge of the Ceiling 188
edgelands 107–111, 118, 119, 174
Edgon Heath 106
Eldorado 86, 188
Elgin, Donald 9
Elidor 82
enchanted landscapes 82
enchantment 6–9, 28, 39, 49, 53, 60, 62, 72, 75, 78, 81, 88–91, 95–98, 100, 101, 103, 119, 144, 170, 177, 189, 199
Ende, Michael 78, 95
English Heritage 179
English Surrealism 177
Enlightenment 7, 11, 28, 87, 88, 92, 121, 123
Entzauberung 7, 87, 89, 100
environmental 9, 17–21, 23, 25, 28, 34, 36, 54, 71, 78, 82, 114, 116, 121, 124, 125, 138, 144, 151, 165, 176, 192, 193, 195–198
Eraserhead 116
Eternal Champion 93, 95, 161
ethics 32, 153, 157
Eurocentric 125, 158, 160, 175
Excalibur 126
experiential research 175

Facebook 182
Faerie Queene, The 85, 142
Fairy folklore 13, 47, 49, 50, 63, 64, 68, 83, 93, 104, 108, 112, 152, 158
Fairyland 11, 54
Fairy Stories 28, 29, 61, 72, 73, 196
Fairy Tale 21, 64, 73, 74, 85, 132, 185
Fairy Tradition 13, 63, 90
fairy way of writing 10, 28, 88, 89

Fallout 120
Fantastical meditations 3, 4
Fantastica (realm) 95
Fantastic Mr Fox 78
Fantastic Voyage 114
Fantastika 19, 94, 107, 123, 135, 148–150, 162, 168, 180, 192, 196
Fantasy and Myth in the Anthropocene 20
Fantasy, definition of 11
Fantasy: how it works 20
Farley, Paul 107, 174
Fellowship of the Ring, The 9
First World assumptions 175
First World War 7, 54, 96, 100, 107, 129, 177, 183
Fisher King 95, 137
Fisher King, The 26, 96, 102, 115
Fisher, Mark 164
Five on a Treasure Island 65
Flâneur 103, 107, 174
Flesh of language 23
flood 28, 30, 31, 48, 68, 74, 116, 142
Folk Horror 125–138, 178, 179
Folk Realism 133, 134, 136, 138
Forbidden Planet 86
Forest Perilous 84, 132
forest bathing 6; *see also* Shinrin-yoku
forests 2, 4, 41, 84, 132
Fourfold Vision 104, 105
fuzzy set 16, 25

Gaia 156, 160, 161
Gaiman, Neil 93, 96, 102, 108, 155, 156, 168, 170
Game of Thrones 3, 124, 161, 191
Garner, Alan 50, 134, 135, 187–189
Gates of Horn, Gates of Ivory 129
Gawain 8, 28, 29, 36–46, 51, 133, 136–138, 190
gentleman's club story 106
Gilchrist, Iain 5, 14
Gilgamesh, Epic of 28, 29, 31, 32, 34, 80, 82, 170, 192
Gilliam, Terry 96
Global North 102, 158, 172
Goblin Market 26, 108
Gods of Pegāna, The 86, 109
Goldberry Riverdaughter 91
Golden Age of Exploration 85, 86
Goldendark 140
Gomorrah 106
Gondal 56, 57
Gondor 56
Google Earth 155
Gormenghast 26, 61, 94, 111, 154
grammar of animacy 101
Grand Map 154, 155, 170
Grand Tour 86
Gray, Alasdair 116
Great Expectations 61
Great North Wood 131
Green Chapel 28, 38–41, 44–46, 138, 190
Green Knight, The (film) 8, 28, 29, 36–41, 43, 45, 46, 136–138, 190
Green Man 37, 46, 71, 126, 131
Greenmantle 112
Grimdark 92, 93, 170
Guardians of the Tall Stones, The 178
Guite, Malcolm 10
Gulliver's Travels 114
Gwynn ap Nudd 183

Hand, Elizabeth 102
Hans Taliesin 126
Haraway, Donna 199
Harrison, M. John 12
Haskell, David George 88
Hawk of May 51, 84
Hemingway's Iceberg 165
Here be dragons 153, 171
Herne the Hunter 8, 76, 77, 126, 131, 132
HEROIC FANTASIES 80
heterotopia 112, 143, 169
High Wild Magic 61
hiraeth 59, 62, 63, 67
Hiroshima 106
His Dark Materials 69, 76, 117, 141, 161
Hobbit, The 54, 65, 73, 76, 164, 183, 184
Hodgson-Burnett, Frances 60, 61, 71
Hogwarts 26, 111, 115, 154
Holdstock, Robert 6, 19
holobionts 199
Holy Grail 85, 95, 96
Homer 70, 80, 81
Honey, I Shrunk the Kids 114
Hope Hodgson, William 107
Hopkinson, Nalo 148, 150, 157
Horse and His Boy, The 78

House of Leaves 143
House on the Borderlands, The 108
How Long 'Til Black Future Month? 118
Howl's Moving Castle 139, 145
Huckleberry Finn 71
Hy Brasil 86

I Am Legend 120
Imaginarium of Doctor Parnassus, The 96, 115
Imaginary Lands
Immanent Grove, The 73, 193
Immrama 70
Industrial Revolution 29, 82, 92, 102
Ingoldsby Legends 72
Inside No.9 127
intersectionality 21, 25, 27
intrusion fantasy 40, 77, 102
IPCC 113, 161, 195
Ireland 9, 48, 50, 66, 83, 108, 175, 191
Island 91

James, M.R. 15
Jansson, Tove 74, 75
Jemisin, N.K 118, 157, 161
Jodrell Bank 188
Jonathan Livingstone Seagull 78
Jonathan Strange and Mr Norrell 142, 147
Jordan College 68, 142, 147
A Journey to Avebury 177
Journey to the West 101
Judge Dredd 116
Juice 113
Jungle Book, The 78

Keats, John 11, 12, 42, 104
Kimmerer, Robin Wall 88, 101, 159, 193
King Arthur 37–39, 90, 103, 137
Kingdoms of Elfin 86, 109
King in Yellow, The 86
King of Elfland's Daughter, The 109
Kings Heath 185
Kirk, Reverend Robert 12
Kuang, R.F. 146, 148, 150

La Belle Dame Sans Merci 104
La Belle Sauvage 68, 76, 141, 142
Lady Chatterley's Lover 62
La Grande Meaulnes 58, 107
Lake Isle of Innisfree, The 175
Lanark 116
Land, The 80–85
Land beyond all dreams, The (poem) 90
Land of the Giants 114
landscape 3, 7, 23, 24, 28, 31, 37, 42, 44, 46, 47, 50, 53, 54, 57, 59, 63, 75–78, 81, 87, 94, 99, 101, 106, 113, 114, 127, 135, 153, 160, 162, 166, 167, 175–177, 179, 181, 182, 186, 188
Lark Ascending 100
Last Homely House, The 91
Last Man 120
Last Man on Earth, The 120
Last march of the Ents 73, 126
Last Unicorn, The 73, 78, 94
Laurelin 91
Lavondyss 71, 99, 100, 128
Le Grande Meaulnes (1912) 58, 107
Le Guin Index, The 196–198
Le Guin, Ursula K. 70, 73, 80, 100, 161, 163, 192–197
leaver-outer 153
Lebenswelt 18, 23, 26
Legolas 91
liminal 11, 34, 55, 58, 107, 144
Lion, The Witch, and the Wardrobe, The 60
Little Grey Men, The 62–65
Little House of Lost Play, The 55
Little John 84
Little Prince, The 70, 71
Little White Bird, The 94
local distinctiveness 132, 165, 166
Loki (TV series) 168
Lolly Willowes 109
London Adventure: an essay in wandering, The 103, 175
London Orbital 107
Loney, The 133, 134
Lord of the Flies, The 65
Lord of the Mines 183
Lord of the Rings, The (LOTR) 8, 37, 61, 73, 77, 80, 81, 91, 93, 124, 126, 164, 183–186
Lost Horizon 91
Lovecraft Country 162
Lovelock, James 156
Lud's Church 190–191

Lud-in-the-Mist 8, 73, 86, 93, 108, 154
Lydney, Romano-British temple 182–183
Lynch, David 116, 117
Lyra Belacqua 26, 68, 71, 76

Mabey, Richard 107
Machen, Arthur 61, 103, 106–107, 174
Mad Max 120
Magical Child 53, 71–72
Magical Cottage 53–59
Magician's Nephew, The 60, 161
Maidens of the Wells 85
Man in the Crowd, The 103
Man Who Killed Don Quixote, The 96
Man Who Was Thursday, The 105
Mappa Mundi 153
Martin, George R.R. 93
Marxism 158
Masefield, John 75
Master and Margarita, The 109
Matheson, Richard 120
Mayfair Witches, The 115
McKean, Dave 117
McKillip, Patricia 144
Meadows, Donella 26
medievalism 158
Medieval Romance 96
Melkor 91
Melmoth the Wanderer 128
Mendlesohn, Farah 9, 12, 16, 25, 29, 60–62, 71, 158
Mercator Projection 153
Merleau-Ponty, Maurice 18, 23, 24
Metamorphosis, The 163
Micronauts, The 114
Middle-Earth 19, 20
midnight disease 14
Midsommar 127, 136
A Midsummer Night's Dream 7
Midwich Cuckoos, The 178
Miéville, China 117, 118
Mimesis 53, 138
Minas Morgul 186
Minas Tirith 186
Ministry for the Future, The 113
Mirkwood 183
Mirrlees, Hope 8
Mischief Acts 8
Mists of Avalon, The 168
Mohamed, Premee 150, 172
Monster Theory 160
A Month in the Country 100
Moominland Winter 78, 157
Moomins 157
Moomins and the Great Flood, The 78, 157
Moominvalley in November 157
Moominworld 114
Moonheart 112
Moon of Gomrath, The 187–189
Moorcock, Michael 93, 100, 161
Moore, Alan 127, 161
Mooreeffocian 105
Mortal Engines 139
Moseley Bog 73, 185, 186
Mount Doom 154
Mycelial Networks 29, 73, 156, 199
Myrddin Wyllt (Wild Merlin) 84
Mythago 99, 129
Mythago Wood 19, 72
Mythago Wood Cycle 6, 7, 58, 98, 100, 128, 158, 161

Nagasaki 106
Name of the Rose, The 145
Nash, Paul 177, 179
National Trust 52, 174, 179, 180, 184, 186, 189
nature 1–7
nature reserve 73, 185, 186, 191
negative capability 11
neoliberalism 25, 194
Neolithic 32, 45, 125, 176, 177
Neverending Story, The 73, 78, 82, 95
Neverland 56, 65, 66
Neverwhere 115
New World 85
New World Fantasy 85
Niamh of the Golden Hair 50, 66, 83, 90
Night Force 112
Nightbreed 115
Nodens 182, 183
Nomad Space 156, 169
nomadic emergence 169
Novik, Naomi 72, 128, 132
'N' (short story by Arthur Machen) 106–107
Númenor 2, 57, 91

Odyssey, The 70, 80
Oisín 50, 66, 83, 90
Okorafor, Nnedi 149, 157, 159, 162, 170
Old Speech 193
Old Straight Track, The 179
Old Wood, The
Ombria in Shadow 144
Omega Man, The 120
On Gallows Down 78
Once and Future King, The
Onyesonwu 71
oral storytelling 20, 28, 36, 72, 139
ordinary world 59, 60
Orlando Furioso 85
Outlaw Varjak Paw, The 117
Outrun, The 136
Ovenden, Graham 177
Oxford 141, 142, 146, 147, 183, 185
Oxford Fantasists 147

Pan 61, 63, 64, 67
paradise 48, 60, 69, 85, 87–101
Parsifal 85, 126, 137
pathetic fallacy 44, 75, 104
Peake, Mervyn 94
Penelopiad, The 168
Peter Pan 61, 66, 69, 71, 94, 174
Phantom Tollbooth, The 60
phenomenological 16–18, 25
phenomenology 18, 21, 23
Philosophy of Composition, The 15
Phoenix 117
Pickering Manuscript, The 104
Piers Plowman 1, 3
Pilgrim 128
Pilgrim's Progress, The 101
Piper at the Gates of Dawn, The (Pink Floyd) 63, 67, 68, 142
Piranesi 61, 142, 143
Placiality 153, 154, 157, 158, 172
Planet of the Apes 120
planetarity 154, 157, 158
plasticity 9, 33, 193, 199
Poe, Edgar Allen 15
Poetics 80
polder 112, 113, 129
Polish folklore 132
Poppy War, The 146, 150
Portal-Quest 58, 60, 61, 95
post-apocalyptic 76, 105, 120–123, 139, 140, 142
postcolonialism 86
Potlatch 103
Power, The 132, 162
Practice of the Wild, The 107
Pratchett, Terry 152
prehistoric 45, 178
prelapsarian 60, 62, 69, 102, 144, 170
Princess Mononoke 71
Prospero's Books 86
prosumer 152, 199
proto-Tolkienian 10, 89
psychogeography 103, 104, 107, 187
Puck of Pook's Hill 61, 62, 64, 71
Puer eternus 51, 71
Pullman, Philip 117, 141, 147, 161, 194
Putter-inner 153, 164
Puzzlewood 180–183

Quatermass: The Final Sequel 116, 178

Raworth, Kate 113
Read, Rupert 196
Realms of Imagination (exhibition) 21, 70
Red Shift 188
re-enchantment 7, 9, 10, 19
Reluctant Messiah 87, 95, 195
Return of the King, The 57
Return of the Native 106
Rice, Anne 102
Riddley Walker 120, 132
Ritual, The 127
Rivers of London 115
Road, The 113
Roadside Picnic 110
Roanhorse, Rebecca 159
Robin Hood 72, 84, 126, 195
Robin of Sherwood 126
Robinson, Kim Stanley 19
Robinsonade 103
Rowling, JK 157
Rule of Names, The 26, 70, 193
Russia 109, 110
Ryhope Wood 6, 7, 58, 98–100, 113, 128–130

Said, S.F. 117, 118
Saint-Exupéry, Antoine de 70

Sangster, Matthew 21
Sapir-Whorf hypothesis 87, 200
Sarehole Mill 73, 185, 186
Scarlet Letter, The 11
Scarp 107
Schama, Simon 1, 153
Science Fantasy 159
seasons 4, 40, 46, 63, 74, 75, 78, 138, 178
Second World War 62, 64, 100
Secondary Worlds 19, 53, 54, 60, 70, 74, 80–82, 86, 95, 132, 142, 144, 146, 148, 167, 173, 188, 193
Secret Commonwealth of Elves, Fauns and Fairies, The 12, 68, 141
Secret Garden, The
A Secular Age 97
Sehnsucht 144, 174
Selfish Giant, The 61
Severn 183
Shakespeare, William 4–7
Shangri-La 91, 111
She 91
Sheldrake, Merlin 20
Shelley, Percy 147
Shepherd, Nan 134, 162, 187
Shinrin-yoku 6, 73
Silbury Hill 177, 179
Silent Spring 78, 93
Silmarillion, The 10
Sir Gawain and the Green Knight 8, 28, 29, 36, 37, 136, 138, 190
Sites of Special Scientific Interest 174, 186
Situationist International 103
Sleep of Reasons Produces Monsters, The 11
Social Dilemma, The 182
social media 182
Society of the Spectacle, The 181
Solnit, Rebecca 156, 162, 167, 174
Somewhere in Time 113
Song of Amergin 48
Song of Ice and Fire 8, 80, 161
Southern Reach series 110
Special World 26, 59, 60
Speculative Fiction 122, 140, 192
Speech of Things, The 77
Spiritwalk 112, 113
Spivak 156, 158
Spufford, Francis 58
Stalker 110, 121
Stardust 73, 93, 108
Star Trek: Voyager 139
Star Wars 124, 180, 181
Station Eleven 139, 193
Stevenson, Robert Louis 13
Stewart, George 121–123
Stig of the Dump 71
Storyworlds 4
Stow-on-the-Wold 184
Strange Case of Dr Jekyll and Mr Hyde, The 13, 14, 103
Strategies of Fantasy 20
Stroud, Allen 21
Subtle Knife, The 60
Summer Book, The 74, 75
Summer Magic/The Journal of Luke Kirby 53, 115
Sunken Land Begins to Rise Again, The 118
Swallows and Amazons 64, 65
Sylvia and Bruno Concluded 154
systems theory 21, 22, 24–27, 29

Taliesin 126
Tam Lin 108
Tamson House 112, 113
Taproot texts 3, 4, 21, 28, 29, 37, 73, 190, 192
TARDIS 111
TCBS 54, 183
Ted Sandyman 185
Tehanu 71
Telperion 91
Tempest, The 4, 85, 86
Temporary Autonomous Zone 169, 179
Tess of the D'Urbervilles 125
Testament of Loki, The 168
Thin places 83
thinning 8, 21, 81, 87, 98
Thomas, Ebony Elizabeth 150
Thor: Ragnarok 168
A Thousand Plateaus 168
Threshold Fantasy 58
Thrutopia 196
Thursbitch 134, 135
Tide Child Trilogy 158
Time Abyss 7, 99; *see also* 'chrono-abyss'
Time Bandits 96, 102

Time Machine, The 99, 129
Tinkerbell Effect 94
Tir nan Og 46, 50, 66, 83, 90, 111
Todd, Barbara Euphan 77, 127
Tom Bombadil 73, 91, 185, 186
Tour, The 70, 85–87, 162
Transliminal moment 12
Treacle Walker 134, 188, 189
Treasure Island 65
Tristan and Isolde 84
Twelve Monkeys 96
Twin Peaks 117
Two Towers, The 126, 186
Two Trees of the Valinor 91

Ultima Thule 15, 86
Under the Pendulum Sun 145, 146, 162
Underground Railroad, The 150, 162
Ungoliant 91
United Nations 113, 196
Uprooted 72, 132
Urban Fantasy 102–123
utopia 85, 111, 144, 156, 169, 196

Varjak Paw 117
Vathek 86
Vesica pisces 55
Victorian Age 7, 15, 29, 47, 92, 96, 103, 122, 125, 145, 155
Vietnam War 194
Viriconium 82, 83, 94
Vyne Ring, The 183

Wainscot societies 21, 68, 74, 114–116
Wakanda 111
Wall Kimmerer, Robin 88, 101, 159, 193
War Hound and the World's Pain, The 95
War in Heaven 107
War of the Worlds, The 106
Warhammer 40,000 92
Warner, Sylvia Townsend 108
Wasteland 51, 81, 83, 95, 106, 110, 126, 140, 170
Water Babies, The 64, 66
Waters of Life 95
Watership Down 78
Watson, Emma 167
Weber, Max 87, 97
Weirdstone of Brisingamen, The 73, 134, 187
What We Do in the Shadows 115
White Ogre, The 185
Whitehead, Colson 150
Who Fears Death 150, 162
Wicker Man, The 125, 127, 133, 135, 136, 178
Wild Twin 32, 131
Wiltshire 45, 177, 178
Wind from Nowhere, The 121
Wind in the Willows, The 61, 63, 64, 67, 142
Windling, Terri 96, 102, 111
Witching Hour, The 115
Withered Hill 135, 136
Wizard's Well 187–189
Womack, Ytasha 149
Wombles, The 114
Woods 4–7
Word for World Is Forest, The 194–196
Word of Unbinding, The 70, 193
World Fantasy Award 100
World Tree 189
Worm Ourobouros, The 86
Worzel Gummidge 77, 78, 127
wrongness 81, 111
Wuthering Heights 56, 57

Y Mabinogi 8, 47, 49, 66, 83, 84, 126

Zoo City 124, 141
Zuni Map Art Project 166, 167

FSC
www.fsc.org
MIX
Papier | Fördert
gute Waldnutzung
FSC® C083411

Kolibri 360 GmbH
Ferdinand-Jühlke-Straße 7
99095 Erfurt, Deutschland
produktsicherheit@kolibri360.de